Charting the Course for a New Air Force Inspection System

Frank Camm • Laura Werber • Julie Kim • Elizabeth Wilke • Rena Rudavsky

Prepared for the United States Air Force
Approved for public release; distribution unlimited

PROJECT AIR FORCE

The research described in this report was sponsored by the United States Air Force under Contract FA7014-06-C-0001. Further information may be obtained from the Strategic Planning Division, Directorate of Plans, Hq USAF.

Library of Congress Cataloging-in-Publication Data is available for this publication.

ISBN: 978-0-8330-7704-2

The RAND Corporation is a nonprofit institution that helps improve policy and decisionmaking through research and analysis. RAND's publications do not necessarily reflect the opinions of its research clients and sponsors.

RAND® is a registered trademark.

© Copyright 2013 RAND Corporation

Permission is given to duplicate this document for personal use only, as long as it is unaltered and complete. Copies may not be duplicated for commercial purposes. Unauthorized posting of RAND documents to a non-RAND website is prohibited. RAND documents are protected under copyright law. For information on reprint and linking permissions, please visit the RAND permissions page (http://www.rand.org/publications/permissions.html).

Published 2013 by the RAND Corporation
1776 Main Street, P.O. Box 2138, Santa Monica, CA 90407-2138
1200 South Hayes Street, Arlington, VA 22202-5050
4570 Fifth Avenue, Suite 600, Pittsburgh, PA 15213-2665
RAND URL: http://www.rand.org/
To order RAND documents or to obtain additional information, contact
Distribution Services: Telephone: (310) 451-7002;
Fax: (310) 451-6915; Email: order@rand.org

Preface

The Air Force Inspection System relies on inspections by the Inspector General (IG) and assessments and evaluations by functional area managers (FAMs) to ensure that all wings comply with Air Force standards and are ready to execute their contingency missions. These oversight activities have grown dramatically over time, despite repeated efforts to limit the burden they place on individual Air Force wings as well as the IGs and the FAMs, all of which are operating under increasingly constrained resources. The Office of the Inspector General of the Air Force (SAF/IG) is now leading an Air Force–wide effort to make significant changes in the inspection system and reduce this burden while at the same time improving the quality of oversight the inspection system provides.

In 2010, SAF/IG asked RAND Project AIR FORCE to support this ongoing effort by collecting new primary data on the inspection system, identifying effective inspection and information collection practices elsewhere that the Air Force might emulate, and providing direct support to SAF/IG and to the cross-functional Headquarters Air Force (HAF) Inspection System Improvement Tiger Team (ISITT) that SAF/IG leads. This document reports the findings of the fiscal year 2011 project "Enhancing SAF/IG's Ability to Meet Its Title 10 Responsibilities."

This document builds directly on past RAND Corporation analysis of the design of performance management and performance-based accountability systems and the implementation of organizational change in various areas of the Department of Defense. Recent examples include:

- Laura H. Baldwin et al., *Strategic Sourcing: Measuring and Managing Performance*, DB-287-AF, 2000
- Frank Camm et al., *Implementing Proactive Environmental Management: Lessons Learned from Best Commercial Practice*, MR-1371-OSD, 2001
- Cynthia R. Cook et al., "Implementation," in Bernard D. Rostker et al., *Sexual Orientation and U.S. Military Personnel Policy: An Update of RAND's 1993 Study*, MG-1056-OSD, 2010, pp. 371–388
- Nancy Y. Moore et al., *Implementing Best Purchasing and Supply Management Practices: Lessons from Innovative Commercial Firms*, DB-334-AF, 2002
- Brian M. Stecher et al., *Toward a Culture of Consequences: Performance-Based Accountability Systems for Public Services*, MG-1019, 2010

This document should interest policymakers and analysts concerned with cost-effective design and the use of formal system-wide oversight mechanisms in large, complex defense activities.

SAF/IG sponsored this research, which was carried out in the Resource Management Program of RAND Project AIR FORCE.

RAND Project AIR FORCE

RAND Project AIR FORCE (PAF), a division of the RAND Corporation, is the U.S. Air Force's federally funded research and development center for studies and analyses. PAF provides the Air Force with independent analyses of policy alternatives affecting the development, employment, combat readiness, and support of current and future air, space, and cyber forces. Research is conducted in four programs: Force Modernization and Employment; Manpower, Personnel, and Training; Resource Management; and Strategy and Doctrine.

Additional information about PAF is available on our website:
http://www.rand.org/paf/

Contents

Preface iii

Figures ix

Tables xi

Summary xiii

Acknowledgments xix

Abbreviations xxi

CHAPTER ONE

Introduction 1

Background on Current and Potential Future Inspections 2

Analytic Approach 5

Roadmap 6

CHAPTER TWO

Choosing a Better Inspection Interval 7

Inspection Frequency and Governance 7

How Air Force Personnel in the Field View the Inspection Interval 11

Scheduling Inspections at the FAA 13

The Design Assessment Phase and Safety Attribute Inspections (SAIs) 14

The Performance Assessment Phase and EPIs 14

Implications for the Air Force 19

Summary 21

CHAPTER THREE

Reducing the Inspection Footprint 23

Synchronize Inspection Events 24

Integrate Synchronized Inspection Events 27

Use Sampling to Reduce the Time Spent to Assess Any Activity During an Inspection Event 30

Sampling and Governance 30

How Air Force Personnel View Sampling in the Field 32

Use a Higher Percentage of No-Notice Inspection Events 36

No-Notice Inspections and Governance 36

How Air Force Personnel View No-Notice Inspections in the Field 37

Summary 39

CHAPTER FOUR
Shift in Relative Emphasis of External Inspection and Wing Self-Reporting 41
Greater Reliance on a Wing's Own Self-Inspection Practices 41
    Perceived Benefits of Greater Reliance on a Unit's Own Self-Inspection Practices 42
    Concerns about This Potential Change 42
    Benefits of an External Look 45
    Suggestions for Implementation 46
    Evidence from Wing Inspection Preparation Activities Suggest That a CCIP Is Feasible 47
Federal Aviation Administration Voluntary Reporting Programs... 48
    Current FAA Voluntary Reporting Programs 48
    Cost Effectiveness of the FAA Voluntary Reporting Programs 52
    Implications for the Air Force Inspection System... 52
    Military Aviation Safety Action Program (M-ASAP)—An Air Force Application 53
Summary.. 54

CHAPTER FIVE
Introducing the New Unit Effectiveness Inspection (UEI) 57
Leadership.. 57
    Definition 57
    Relationship with Performance 58
Discipline 62
    Definition 62
    Relationship with Performance 62
Measurement of Leadership and Discipline... 64
    Approaches Suggested by the Literature 64
    Approaches Suggested by Current Air Force Practice 65
Concerns Regarding the Assessment of Leadership.. 72
Summary.. 74

CHAPTER SIX
Introducing the Management Internal Control Toolset (MICT).. 77
Pros and Cons of a Standardized Wing-Level IT System 78
    Potential Benefits... 78
    Potential Challenges... 79
Will MICT Benefit Local Users More Than It Costs Them?... 80
Will Wing Personnel Use MICT Appropriately?... 82
Sharing Data from MICT Outside the Wing 84
Summary.. 86

CHAPTER SEVEN
Implementation of Significant Change in the Inspection System 87
Recent Formal Change Management Perspectives in the United States 87
    Element 1: Plan 88
    Element 2: Execute 91
    Element 3: Sustain... 93

A Case Study: The Evolution of Policy and Implementation of Change in the Federal Aviation
    Administration Inspection System..95
  The Creation of the Aviation Safety Reporting Program (ASRP)........................95
  Shift in Aviation Safety Culture in the Mid-1980s with the Post-Deregulation Traffic Surge 96
  Shift in Aviation Safety Culture in the Mid-1990s with Increased Airline Responsibility.......... 97
  Significant Investment in Database Systems...98
Summary..98

CHAPTER EIGHT
Recommendations.. 101
General Recommendations.. 101
Selecting a Better Inspection Interval 103
Reducing the Inspection Footprint 104
Increasing the Emphasis on Self-Inspections and Self-Reporting......................... 105
Introducing the New Unit Effectiveness Inspection 106
Introducing the Management Internal Control Toolset................................... 107
Implementing Significant Organizational Change 108
Conducting Additional Analysis to Support Implementation............................ 111

APPENDIXES
A. Analysis of Practices the Air Force Inspection System Might Emulate....................... 113
B. Analysis of the Experiences of Air Force Personnel in the Field.......................... 117
C. Risk Management in the Federal Aviation Administration (FAA) Inspection System 127
D. Additional Background on the Air Force Climate Survey 131
E. Additional Background on the Air Force Culture Assessment Safety Tool (AFCAST)..... 137

Bibliography.. 145

Figures

1.1. Future Air Force Inspection System, as Envisioned Fall 2011................................ 4
C.1. Effects of Likelihood and Severity on Level of Risk.. 129

Tables

2.1. Operational Systems, Subsystems, and Elements Relevant to Air Worthiness 16
2.2. Air Carrier Assessment Tool (ACAT) Risk Indicators 18
2.3. Outsource Risk-Scoring Scheme Under Environmental Criticality (EC-04) 20
3.1. Potential Effects of Efforts to Reduce the Inspection System Footprint 25
5.1. Evidence from Fieldwork on Defining Leadership 59
5.2. Evidence from Fieldwork on the Leadership-Performance Link 60
5.3. Evidence from Fieldwork on Defining Discipline 63
5.4. Dimensions of Leadership Measured in MLQ 65
5.5. Edmondson's Measures of Psychological Safety and Team Leader Coaching 66
5.6. Air Force Climate Survey Measures of Direct Supervisor 67
5.7. Air Force Climate Survey Measures of Unit Commander or Commander
      Equivalent 68
5.8. Air Force Climate Survey Additional Measures of Unit Commander or Commander
      Equivalent 69
5.9. Air Force Climate Survey Additional Measures of the Unit's Chain of Command 69
5.10. AFCAST Questions About Supervision 71
7.1. Elements of Formal Change Management Emphasized in Recent Publications 89
E.1. AFCAST Survey Questions for Operations, Maintenance, Support, and Higher
      Headquarters 141

Summary

Air Force senior leadership relies on inspections, assessments, and evaluations to advise it on the efficiency, effectiveness, readiness, and level of compliance of Air Force activities. Conducted by several different organizations within the Air Force, these oversight tasks have grown dramatically over time, despite repeated efforts to limit the burden they place on individual Air Force units. Although Office of the Inspector General of the Air Force (SAF/IG) inspections constitute only about one-fourth of this burden, SAF/IG has the responsibility to set inspection policy and oversee the inspection and evaluation systems for the Air Force as a whole. In 2010, SAF/IG began an aggressive effort to improve inspection policy by reducing the burden it places on inspected units and increasing the quality of relevant information it generates for the Secretary and Chief of Staff of the Air Force and for commanders throughout the Air Force. At SAF/IG's request, in late 2010, RAND Project AIR FORCE (PAF) joined this effort. The RAND Corporation conducted analyses related to five major inspection system goals that SAF/IG and its Inspection System Improvement Tiger Team (ISITT) were pursuing:

1. choosing a better inspection interval
2. reducing the inspection footprint
3. increasing the emphasis on self-inspections and self-reporting
4. introducing the new Unit Effectiveness Inspection (UEI)
5. introducing the Management Internal Control Toolset (MICT).

We relied on multiple data sources to inform our analysis: a review of practices the Air Force Inspection System might emulate, such as the Air Force Culture Assessment Tool program (AFCAST), the Air Force Climate Survey, and the Federal Aviation Administration (FAA) inspection system; an investigation of Air Force personnel's experiences in the field, which included the observation of a compliance inspection, focus groups with inspectors, and interviews and focus groups with members of recently inspected wings; and a review of literature on leadership, organizational change, and other topics.

Choosing a Better Inspection Interval

Under the existing inspection system, the interval between inspections varies significantly across the Air Force. The Air Force leadership is moving toward standardizing and shortening this interval so that one major inspection occurs at every non-nuclear, active component wing every two years. This will ensure that wing commanders, who usually serve in a wing for two years, will face a major inspection during each tour. When we asked Air Force inspectors and inspectees about this proposed change, more favored this approach than any other.

In contrast, the FAA uses a very different approach to inspect the commercial aviation system. It carefully varies the inspection interval across activities to reflect (1) the inherent risk associated with specific aviation-related activities; (2) attributes unique to individual inspected organizations, like the history of their past performance or the stability of their operating environment or leadership team; and (3) the resources available to conduct inspections in any year. We suspect this tailored approach allows the FAA to use its constrained inspection resources to reduce more risk than would the proposed Air Force approach, given the same resources. Some Air Force personnel suggested that an inspection frequency consistent with the FAA's approach may be feasible for the Air Force as well.

Reducing the Inspection Footprint

The Air Force is seeking ways to reduce (1) the number of days each year that a wing is subject to some external oversight event, and (2) the resources consumed—by both inspectors and inspectees—for each event. To this end, the Air Force has already begun synchronizing IG inspections and functional assessments so that they occur on the same days. It also plans to integrate such events so that fewer external inspectors and assessors are required and wing personnel spend less time preparing for and talking with those who come. The Air Force inspectors and inspectees we spoke to generally support such an approach, but also noted that achieving effective integration will be challenging.

The FAA relies heavily on both formal sampling and no-notice inspections in its own inspection system. The Air Force could use formal sampling strategies, which require less information than is typically collected under the existing system, to assess wing performance. The Air Force could also make greater use of no-notice inspections—inspections that occur with very little advance notice—to keep wings on their toes at all times and reduce the resources required to complete individual inspections. The Air Force inspectors and inspectees we talked to do not have strong views on sampling, but generally favored much greater use of no-notice inspections.

Increasing the Emphasis on Self-Inspections and Self-Reporting

Our investigation of Air Force personnel's experience in the field revealed concerns about relying more heavily on the wings' self-inspection programs, at least in their current incarnation. Inspectors noted that both the quality and the nature of self-inspection programs varied greatly under the existing inspection system. Also, personnel from recently inspected wings felt that some units lack people with the skills required to detect and resolve weaknesses without external support. Both inspectors and inspectees worried that personnel within a wing find it hard to be honest with themselves about the weaknesses of that wing and resist reporting such weaknesses outside the wing.

However, the FAA's successful voluntary reporting programs can serve as models the Air Force might consider to encourage honest self-reporting by wing personnel. In addition, personnel themselves proposed a number of ways to place greater emphasis on self-inspection and self-reporting work, such as having wing personnel provide an "external look" at other parts of the wing. The FAA's experience and the ideas shared by Air Force personnel themselves hint that SAF/IG's plan to develop a more robust commander's inspection program (CCIP) is feasible. Moreover, actions some wings already take to prepare for compliance inspections, such as internal Tiger Teams and compliance exercises, suggest that requiring wings to use a variety of formal self-inspection mechanisms may not be a drastic departure from current practice.

Introducing the New Unit Effectiveness Inspection (UEI)

SAF/IG's vision for a new inspection system includes a new type of inspection, the UEI, which will subsume elements of the compliance inspection in place at the time of our research. One planned component of the UEI is an assessment of a wing's discipline and leadership. Our fieldwork and review of scholarly research did not yield definitive guidance on how to measure discipline. In contrast, however, measuring leadership within the new UEI holds promise.

Overall, our results indicate that, in spite of some inspectors' reluctance to assess leadership during a compliance-focused inspection (based on their experience with the existing system), there are both the reason and the means to do so. Specifically, we found compelling evidence of a link between leadership and performance, and identified several well-validated, practicable ways to measure leadership. Air Force personnel cited a number of leadership characteristics they deemed important that have already been operationalized in the Air Force Manpower Agency's Climate Survey and AFCAST survey, for instance. Methods developed by academics, such as the Multifactor Leadership Questionnaire (MLQ), psychological safety measures, and the use of data aggregation, suggest additional and often complementary ways of assessing not only the effectiveness of a wing commander, but also that of the entire wing leadership chain of command.

Introducing the Management Internal Control Toolset (MICT)

In 2009, the Air Force began to introduce a new standard information management tool to all of its reserve component wings. MICT allows a wing to record and manage information on items from an inspection checklist or any other item the wing commander deems important to the wing's performance. When this information reveals a shortfall, MICT facilitates the management of a corrective action program that tracks progress until the root cause of the shortfall is brought under control. Based on overall positive experience to date, the Air Force is now introducing MICT to all active component wings.

The Air Force inspectors and inspectees we spoke to generally like the idea of a system like MICT, but are quite skeptical that MICT will yield the benefits it has promised. Their past experience with new information systems has led them to worry that MICT will be too hard to use, will not work as well as the local systems they use now, will not come with enough resources to sustain appropriate training and user support, and might even invite resource managers to cut resources for inspectors based on the belief that fewer inspectors will be needed after MICT is introduced. There is no objective evidence from Air Force experience to date to support these concerns, but the Air Force will need to address them to ensure the success of its MICT implementation.

Implementing Significant Organizational Change

The changes the Air Force leadership is pursuing raise basic cultural issues that must be addressed before these changes achieve their maximum benefits. A formal approach to change management has emerged and evolved over the last three decades that is well suited to facilitating this kind of change in a large, complex organization like the Air Force.

This new approach to formal change management (1) plans for a change, (2) then executes it, and (3) finally, sustains the change until it becomes part of routine operations. The planning process addresses the tightly interrelated tasks of designing change, creating high-level support, and convincing individual organizational members that they will benefit more from the change than from opposing it. It breaks a change into manageable chunks, like blocks of new aircraft, each of which can be built relatively quickly. The execution stage uses training, monitoring, adjustment, and extensive communication to learn from the ongoing implementation of each incremental chunk, correct weaknesses quickly as they are exposed, and provide senior leaders with constant empirical evidence that the change is yielding its expected benefits. Sustainment migrates each incremental change to the dominant command and control system of the organization as a whole.

The FAA has used elements of this approach to effectively achieve changes similar to those the Air Force is now seeking to implement. The Air Force can particularly learn from how the FAA has pursued major change in its own inspection system since the 1970s. In this report, we offer many possible ways of doing this, as summarized below.

Recommendations

Through our own analysis and discussions with Air Force personnel in SAF/IG and the ISITT, we formulated the following recommendations, which are presented in detail in Chapter Eight of this report.

General Risk Management

Consider adopting a formal risk management system to guide Air Force inspection-related decisions and activities. SAF/IG should take the lead in developing a risk management system suited to the new inspection system. Without such a system in place, several of the recommendations below (marked with asterisks) may not be feasible.

The Inspection Interval

Initially condition the frequency of inspection of specific activities on risk management factors.* Over the longer term, revisit the decision to move to one major inspection every two years for each wing.*

The Inspection Footprint

As future external inspections become more focused and reduce in size, ensure that they continue to capture the priorities of the IG and the relevant functional area communities.* Apply formal sampling guidance to reduce the burden of inspections and increase their productivity.* Use information on a wing's past performance to design the focus and depth of each full inspection.

Self-Inspection and Self-Reporting

Foster conditions for psychological safety to increase the willingness of all individuals in wings to report weaknesses. Consider adapting some aspects of FAA's voluntary reporting system as part of the new CCIP. This system would maintain the anonymity of individuals reporting from within wings to encourage more honest reporting.

Support wings' efforts to preserve the "external look," a mechanism many wings use today to have personnel assess each other across squadrons or other wings.

Measures of Leadership

Ensure that measures of leadership take into consideration the full chain of command, not just the wing commander. Recognize that leadership and discipline are multi-faceted constructs and measure them as such. Consider the use of qualitative methods to measure leadership, but ensure that they are standardized across inspection teams and sites.

Develop a new UEI survey that adopts items from existing survey instruments. Use other existing data sources to inform the inspection process.

MICT

Follow through to ensure that MICT is implemented cost-effectively. Institute and sustain an approach to using MICT that maintains (1) standard core information and (2) wing-unique information. Recognize MICT as a complement to external inspections and assessments and internal self-inspection, not a replacement for them. Maintain the accuracy of any information in MICT that is freely available to external overseers at the major command (MAJCOM) or Air Force level.

Full, Air Force–Wide Implementation

As the inspection system changes, keep in mind that it has many moving parts and operates as part of a broader governance system. Anticipate and disarm negative perceptions about proposed changes. Expect that full implementation will take time and plan for this by breaking change into incremental chunks and managing each end-to-end. Use formal pilot tests to help monitor and refine increments of change before they are implemented throughout the Air Force.

Additional Analysis to Support Implementation

Develop more detailed quantitative analysis of the costs of the inspection system. Translate the risk assessment *framework* recommended here into guidance for an Air Force risk assessment *system*. Develop concrete and specific guidance that translates formal sampling methods into practical instructions inspectors can apply to increase the quality of information they can collect with given resources. Develop the basis for a more precise and operational definition of discipline. Tailor the broad implementation guidance offered in Chapter Seven to an Air Force setting.

Acknowledgments

We thank Lt Gen Marc E. Rogers, who sponsored this research and took a personal interest in it throughout its execution. In his office, Col Robert D. Hyde worked with us as our principal point of contact and a constant source of good ideas for potential directions in which to take our analysis. Lt Col Lori J. Stender managed myriad details of information exchange and meeting coordination. CMSgt Johnny L. Collett worked closely with us to set up focus groups and executive-level interviews in the field.

Throughout this project, we participated in meetings of the ISITT that Col Hyde chaired. The discussions in those meetings were invaluable to our analysis and often led to more in-depth exchanges with individual members.

We benefited greatly from focus groups held with MAJCOM IG personnel and from interviews and focus groups with personnel from wings that had completed a compliance inspection in the spring/summer 2011 time frame. Because we wish to keep confidential the names of personnel who participated in those focus groups and interviews as well as the specific wings we included in our study, we cannot acknowledge our Air Force facilitators by name, but we recognize the tremendous effort of those individuals who organized and scheduled these focus groups and interviews, and are especially grateful to the many Air Force inspectors and wing personnel who participated candidly in them.

We also conducted many interviews elsewhere that we can acknowledge without compromising the identities of Air Force personnel who gave us their personal views on the inspection system. We thank Col Gregory A. Myers, Air Force Inspection Agency; Col Thomas A. Bussiere, Air Force Global Strike Command (AFGSC/IG); Col Warren Thomas, Air Mobility Command; Lt Col Thomas Hughes, Kathleen Armstrong, and Kevin Tibbs, Air Force Safety Office; Maj Shanon Anderson, 305 AMW/XP, McGuire Air Force Base, N.J.; Maj Heather L. Morgenstern, USAFR/IGIA; James Cross, FAA Flight Standards National Field Office; David Gilliom, FAA Aviation Safety Flight Standards Service; Tony Fazio, FAA Office of Accident Investigation and Prevention; Linda J. Connell, NASA Human Systems Integration Division; and Dr. Anthony P. Ciavarelli, Human Factors Associates, Inc.

Brig Gen Michael A. Longoria (ret.) offered useful insights and references in a discussion and formal peer review.

We also benefited from the contributions of many RAND colleagues. Meg Harrell and Alexandria Felton assisted the authors with the fieldwork for this project. Michael Greenberg provided a thoughtful formal peer review. Kristin Leuschner assisted in the dissemination of the findings. Hosay Yaqub and Donna White provided administrative support, and Donna White also played a key role in document production. We want to give a special thanks to Col Daniel F. Merry, who participated in this project while stationed at RAND as an Air Force fellow. He brought us invaluable, recent experience with and insight into the elements of the Air Force inspection system that Lt Gen Rogers and the ISITT wanted to address.[1]

We thank them all, but retain full responsibility for the objectivity, accuracy, and analytic integrity of the work presented here.

---

[1]   The ranks and offices of all those listed in this section are current as of the time of the research.

Abbreviations

ACAT	Air Carrier Assessment Tool
ACC	Air Combat Command
AETC	Air Education and Training Command
AFCAST	Air Force Culture Assessment Safety Tool
AFI	Air Force Instruction
AFIA	Air Force Inspection Agency
AFMA	Air Force Manpower Agency
AFRC	Air Force Reserve Command
AFS021	Air Force Smart Operations for the 21st Century
AFSC	Air Force Safety Center
AMC	Air Mobility Command
AQP	Advanced Quality Program
ASAIS	Aviation Safety Analysis Information Sharing
ASAP	Aviation Safety Action Program
ASRP	Aviation Safety Reporting Program
ASRS	Aviation Safety Reporting System
ATOS	Air Transportation Oversight System
CASS	Continuing Analysis and Surveillance
CC	Commander
CCIP	Commander's Inspection Program
CCRI	Command Cyber Readiness Inspection
CD	Air Carrier Dynamics
CEO	Chief Executive Officer
CI	Compliance Inspection
CMO	Certificate Management Office
CMT	Certificate Management Team
COTS	Commercial Off-the-Shelf
DA	Design Assessment
DoD	Department of Defense

EC	Environmental Criticality
EPI	Element Performance Inspection
ERC	Events Review Committee
ESOHCAMP	Environmental, Safety, Occupational Health, Compliance Assessment, and Management Program
FAA	Federal Aviation Administration
FAM	Functional Area Manager
FOIA	Freedom of Information Act
FOQA	Flight Operation Quality Assurance
HAF	Headquarters Air Force
HFACS	Human Factors Analysis and Classification System
HHQ	Higher Headquarters
HRO	High Reliability Organization
HSI	Health Services Inspection
IG	Inspector General
ISITT	Inspection System Improvement Tiger Team
ISO	International Organization for Standardization
IT	Information Technology
LCAP	Logistics Compliance Assessment Program
LOSA	Line-Oriented Safety Auditing
M-ASAP	Military Aviation Safety Action Program
M-FOQA	Military Flight Operations Quality Assurance
MAJCOM	Major Command
MEL	Minimum Equipment List
MGA	Major Graded Area
MICT	Management Internal Control Toolset
MLQ	Multifactor Leadership Questionnaire
MOC	Mission Operations Center
MOSE	Model of Organizational Safety Effectiveness
MSG	Mission Support Group
MUNS	Munitions Squadron
Mx	Maintenance
NASA	National Aeronautics and Space Administration
NCO	Non-Commissioned Officer
Ops	Operations
ORI	Operational Readiness Inspection
OS	Operational Stability
OSHA	Occupational Safety and Health Administration

PA	Performance Assessment
PAF	Project AIR FORCE
PH	Performance History
PI	Principal Inspector
POC	Point of Contact
PRA	Probabilistic Risk Assessment
QA	Quality Assurance
QAE	Quality Assurance Evaluator
RMP	Risk Management Process
SAF/IG	Office of the Inspector General of the Air Force
SAI	Safety Attribute Inspection
SAS	Safety Assurance System
SIP	Self-Inspection Program
SME	Subject Matter Expert
SMS	Safety Management System
SPAS	Safety Performance Analysis System
Sup	Support
TIG	The Inspector General of the Air Force
UCI	Unit Compliance Inspection
UEI	Unit Effectiveness Inspection
UIF	Unfavorable Information File
U.S.C.	United States Code
VDRP	Voluntary Disclosure Reporting Program

CHAPTER ONE

Introduction

The Inspector General of the Air Force (SAF/IG), "when directed by the Secretary or the Chief of Staff" of the Air Force, has the responsibility to "inquire into and report upon the discipline, efficiency, and economy of the Air Force" and the Air National Guard.[1] SAF/IG shall also "perform any other duties prescribed by the Secretary or the Chief of Staff." SAF/IG interprets this role from the historical perspective of inspectors general going back at least to Napoleon. SAF/IG has the authority and responsibility to act as the Secretary and Chief of Staff's "eyes and ears," seeking out and monitoring the information they would look for if they had the time and resources to do so themselves.

The senior leaders of the Air Force rely on inspections, assessments, and evaluations to collect information on the status of activities throughout the Air Force. Inspectors general (IGs) in major command (MAJCOM) headquarters schedule and conduct regular inspections of their subordinate units. The Air Force Inspection Agency (AFIA) conducts some additional inspections directly for SAF/IG. Functional area managers (FAMs) in MAJCOM headquarters schedule and conduct assessments and evaluations of subordinate units: some mandatory, some at the request of these units. Activities outside the Air Force, including the Government Accountability Office (GAO), the Department of Defense (DoD) IG, the Defense Contract Management Agency, and various hospital, university, and prison certification boards, also conduct their own oversight events with Air Force units.

These oversight activities have grown dramatically over time, despite repeated efforts to limit the burden they place on individual Air Force units. In 1947, the Air Force conducted just six types of inspections. By 2010, this number had grown to over 97 types of inspections, assessments, and evaluations. In 2009, a wing commander could expect to have some external oversight event occurring on 57 percent of the days of the year, leaving only 43 percent of the days available as "white space," or time during which the commander could focus solely on the wing mission (Rogers, 2010).

While SAF/IG inspections only account for about 28 percent of the external oversight burden on wings,[2] the Secretary and Chief of Staff have given SAF/IG the responsibility to set inspection policy and oversee the inspection and evaluation systems for the Air Force as a whole. Acting in this role, SAF/IG began an aggressive effort in 2010 to improve inspection

---

[1] The quotations in the text are from 10 U.S.C., Sec. 8020. 32 U.S.C., Sec. 105 defines analogous roles with regard to the Air National Guard.

[2] Rogers, 2010. This is measured in terms of inspector/assessor man-days at a wing, adjusted for the average interval between oversight events. The Air Force has no primary measure of oversight burden. Note that this measure, which focuses on the costs of inspector/assessors, is qualitatively different from the measure of white space, which focuses on the costs experienced by units subject to oversight.

policy by reducing the burden that policy places on inspected units and increasing the quality of relevant information it generates for the Secretary and Chief of Staff of the Air Force, and for commanders throughout the Air Force. At SAF/IG's request, RAND Project AIR FORCE (PAF) joined this effort in late 2010. This report documents PAF's analytic findings on the issues that SAF/IG asked PAF to examine.

SAF/IG created an Inspection System Improvement Tiger Team (ISITT), composed primarily of colonels and their civilian counterparts, responsible for inspection, assessment, and evaluation policymaking across the functional areas in Headquarters Air Force (HAF). This group met frequently from December 2010 through December 2011, during which time it actively debated different ways of improving the inspection system, refined an approach that the senior Air Force leadership accepted, and began the arduous process of "socializing" this approach across the functional communities and MAJCOMs of the Air Force. We, the PAF study team, conducted analysis alongside the ongoing work of the ISITT, collecting information we believed would be useful to SAF/IG and the ISITT and correlating our focus with that of SAF/IG and the ISITT. We regularly briefed this group on our interim findings and drew on the expertise of its members to improve our understanding of the Air Force inspection system.

Background on Current and Potential Future Inspections

Given the primary policy changes under consideration, described below, SAF/IG asked PAF to focus its analysis on compliance inspections (CIs) at major wings.[3] According to the Department of the Air Force at the time of this analysis (2009, p. 25), CIs were conducted (1) to evaluate adherence to public law, executive orders, and DoD, Air Force, and MAJCOM directives and instructions; and (2) to assess areas of operations critical to mission success. The Air Force used CIs not to specifically rate organizations, but to rate programmatic compliance.[4] Under

---

[3] A wing is an organization within a MAJCOM with a "distinct mission with significant scope. A wing is composed of a primary mission group (e.g., operations, training) and the necessary supporting groups," typically maintenance, mission support, and medical groups. Wings have a minimum adjusted population of at least 1,000 (Air Force Instruction [AFI] 38-101, 2012a, pp. 12, 22–23) and are commanded by a colonel or brigadier general. The largest and most complex CIs occur at wings; ongoing changes in Air Force policy tend to emphasize inspections at wings. CIs are one of four major types of Air Force IG inspections that occur at wings. Operational Readiness Inspections (ORIs) "evaluate and measure the ability of units to perform their wartime, contingency, or force sustainment missions." Nuclear Surety Inspections (NSIs) are "performance and compliance-based inspections and are conducted to evaluate a unit's ability to manage nuclear resources while complying with all nuclear surety standards." Nuclear Operational Readiness Inspections (NORIs) are "performance-based readiness evaluation[s] of nuclear-tasked units which support United States Strategic Command . . . and Joint Chiefs of Staff" operational plans (AFI 90-201, 2009, pp. 37, 46, 64). Wings were also subject to a wide variety of smaller functional assessments. Given project resources and the intended order of Air Force inspection system changes at the time this study began, we focused our efforts on CIs within the active component Air Force. Changes to the reserve component Air Force inspection system are under consideration as well.

[4] AFI 90-201 was updated in March 2012. Policy now states that Active Duty Unit "commanders at wing level and below should receive at least one major IG inspection during their command tour . . . CUI [Combined Unit Inspection] Phase 0, 1, and 2 requirements must be completed within 48 months, while striving to complete these requirements within 24 months" (AFI 90-201, 2012b, p. 19).

that system, CIs of wings or wing-level equivalents were conducted at intervals of no longer than 60 months.[5]

The amount of notification given prior to a full-scale CI is determined by the MAJCOM IG, but typically is about one year. As the inspection date approaches, an inspection team comprising IG personnel and, for some MAJCOMs, functional augmentees, interacts with personnel from the wing that is to be inspected to prepare for the team's visit. This interaction may include requests for documentation as well as logistical coordination. Depending on the size of the wing, the inspection team may be as large as 100 people, and they require billeting, transportation, food, and orientation materials.

Typically, at an active component wing, the inspection team spends approximately one week on site. Members of the inspection team break off into small groups to inspect distinct Major Graded Areas (MGAs), returning at the end of each inspection day to provide the inspection team chief with a progress update, to air any emerging concerns, and to participate in team meetings as the team chief deems appropriate. The team chief and his or her deputy apprise the wing command of the team's progress and serve as liaisons between the wing and the IG's inspection team. At the conclusion of the inspection period, the team gathers together to write its inspection report and assign compliance-oriented grades. Depending on the MAJCOM, the extent of compliance may be reported using either a three-tiered rating system (does not comply, complies with comments, complies) or a five-tiered rating system (unsatisfactory, marginal, satisfactory, excellent, outstanding). Additional feedback, such as best practices and outstanding individual performers, may also be included. The report is distributed to the commander of the inspected wing, the MAJCOM IG, and the MAJCOM commander. If deficiencies have been noted, the wing is expected to take and document corrective actions.

Overall, CIs are intended to serve as an important part of the Air Force's inspection system. They provide an external, unbiased look at day-to-day compliance that complements operational readiness inspections and, if a wing has a nuclear mission, nuclear surety inspections.

During 2010 and 2011, SAF/IG and the ISITT considered many alternatives to the current approach to inspections and ultimately emphasized the following five major ideas for change:

1. Ensure that a major inspection occurs during each tour of senior leadership at every wing.
2. Given the increased frequency of inspections, reduce the burden they place on wings by reducing their footprint—the time and resources consumed during external oversight events—and increasing white space—time free from external oversight events—at the wing.
3. Increase the capabilities and motivation of wings to conduct regular, standardized, rigorous self-inspections that SAF/IG can rely on to identify and correct most problems before they are detected by an external oversight team.

---

[5] The actual interval varied across MAJCOMs and often across time within MAJCOMs. Some intervals were set by higher-level Air Force policy. Most were chosen by MAJCOMs, based on striking an appropriate balance between operational risks and resources available within any particular MAJCOM. We do not address such variations, which SAF/IG is already well aware of.

4. Introduce a new Unit Effectiveness Inspection (UEI) that collects information more relevant to the senior leadership of the Air Force while requiring fewer resources, both of the wing and the external oversight team.
5. Standardize the use of the Management Internal Control Toolset (MICT) throughout the Air Force.

These ideas are reflected in the new inspection system envisioned at the time of this report's writing. As shown in Figure 1.1, this prospective system calls for accountable, structured internal inspection, achieved largely through the new multi-faceted Commander's Inspection Program (CCIP) and enabled by MICT, an IT-based tool that helps execute aspects of the CCIP and produces results visible to higher headquarters staff. A wing's internal inspection is, in turn, verified by IG-led external inspections, including a new UEI that assesses leadership effectiveness, military discipline, and aspects of wing climate and culture. MAJCOM staff provide support as needed via policy, guidance, training, and resource recommendations. Taken together, the IG-based inspections and verification, wing-level CCIP, and MAJCOM staff inputs will provide wing commanders and those at higher levels of command with a robust picture of wing functionality through a variety of metrics, including efficiency, effectiveness, and readiness. Specifically, this new system is intended to answer questions that Lt Gen Marc Rogers, the Inspector General of the Air Force, referred to as the "Big 7" (Rogers, 2011):

1. Are units properly manned?
2. Are units properly trained?
3. Are units properly equipped?

Figure 1.1
Future Air Force Inspection System, as Envisioned Fall 2011

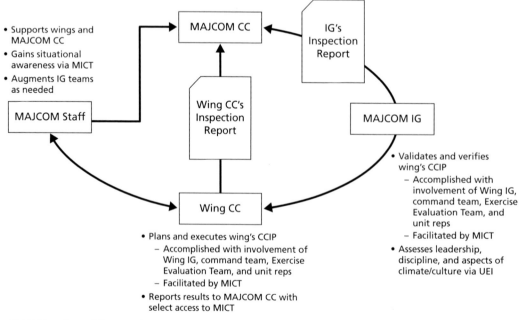

SOURCE: Adapted from Hyde, 2011b.
RAND TR1291-1.1

4. What's the condition of unit equipment?
5. Are units ready and proficient?
6. Is the leadership climate effective?
7. Are units disciplined and compliant?

Analytic Approach

We relied on multiple data sources to inform our analysis: a review of practices the Air Force inspection system might emulate, an investigation of Air Force personnel's experiences in the field, and a literature review. For our review of practices to emulate, we evaluated practices both inside and outside the Air Force in order to glean the most relevant and useful lessons for SAF/IG. A description of how we approached such practices is provided in Appendix A. Ultimately, we focused on the Air Force Culture Assessment Safety Tool (AFCAST) program, the Air Force Climate Survey, and the Federal Aviation Administration (FAA) inspection system.[6]

Our investigation of Air Force experiences in the field consisted of three parts. At the outset, in spring 2011, three members of the PAF project team observed a CI firsthand, including the inspection kick-off meeting, team meetings, and discussions with wing command. We also "rode along" with inspectors as they inspected MGAs. In addition, from April through June 2011, we conducted focus groups with inspection team members from three MAJCOMs: Air Combat Command (ACC), Air Education and Training Command (AETC), and Air Mobility Command (AMC).[7] Nine focus groups were conducted in total, three per MAJCOM, and 71 IG personnel and functional augmentees participated overall. Topics discussed included inspection preparation activities, the on-site inspection process, the use of sampling, leadership, discipline, and potential changes to the inspection system. Finally, we visited three wings, one from each of the aforementioned MAJCOMs, soon after they underwent a CI (June–July 2011). At each wing, we conducted interviews with wing and group-level leaders, and focus groups with wing personnel. In total, we completed 12 interviews with 27 unit leaders (higher-level officers and non-commissioned officers [NCOs]), and nine focus groups with 69 wing personnel (officers and NCOs). During these sessions, we discussed inspection preparation activities, perceptions of the recent CI experience, leadership, discipline, self-inspection, error reporting, and potential changes to the inspection system. More details about our fieldwork,

---

[6] SAF/IG explicitly asked us to examine AFCAST and the FAA inspection system. We added the Climate Survey because it appeared to offer an existing data source that SAF/IG could use to address questions relevant to the potential policy changes discussed here. Very briefly, AFCAST and the Climate Survey offer specific ways to measure aspects of the quality of leadership in a wing. The FAA offers potentially useful insights into how to (1) make large changes in the culture of an aviation-focused inspection system, (2) assess risks relevant to the design of a cost-effective inspection system, and (3) improve the quality and reliability of data collected through the inspection system. SAF/IG had an immediate interest in all of these topics.

[7] We worked with the sponsor to choose these three MAJCOMs to ensure that the study looked across different mission types and variations in inspections likely to result from these different mission types. The study is not designed to dive into details about individual MAJCOMs, a task that would require far more resources than were available for this study. Changes in MAJCOM-specific inspection practices will have to be sensitive to large differences in the missions that the inspection system examines across the Air Force. The Air Force is collecting more information on such differences as the implementation of inspection reform continues.

including how units and individuals were selected, interview and focus group protocols, and our analytic approach, are provided in Appendix B.

Finally, we conducted a review of scholarly literature, largely from management and psychology disciplines, to complement our review of practices and our investigation of Air Force experiences in the field. During the course of our project, we reviewed publications on generational diversity, behavior change, information sharing, organizational culture, leadership, psychological safety, and managing organizational change. The latter three topics figure prominently in this report.

Our analysis often discusses the "burden" or cost imposed by the inspection system as well as ways to make the inspection system more cost-effective. In this context, it might seem natural to measure the cost of the current inspection system. However, given the resources available for this study, our sponsor asked us to focus on other analytic tasks. More detailed information about the costs and benefits of certain elements of the inspection system should allow for more specific recommendations than those we offer here. As the Air Force continues to implement change and refine its concepts for a new inspection system, the collection of more precise cost information would likely be useful.

Roadmap

The next four chapters present our findings on four separate inspection system goals: aligning the inspection interval to the tour length of wing leaders (Chapter Two), reducing the burden of inspections on wings and inspectors (Chapter Three), shifting the relative emphasis on wing self-reporting (Chapter Four), and introducing the new UEI (Chapter Five). In Chapter Six, we address a key enabler of the envisioned inspection system, MICT. Chapter Seven includes a discussion of key findings from organizational change research as well as a case study of the FAA's management of a related change. Last, in Chapter Eight, we draw on the analytic findings reported in previous chapters to offer a series of policy recommendations.

Appendices A and B present additional information about how we structured and interpreted our review of practices to emulate and our investigation of Air Force personnel's experiences in the field, respectively. Appendix C provides additional details on the role of risk management in the FAA inspection system. Appendices D and E provide additional details on two survey programs that the Air Force inspection system might emulate or use to complement its own data collection systems: the Air Force Climate Survey (Appendix D) and the AFCAST (Appendix E).

CHAPTER TWO

Choosing a Better Inspection Interval

The Air Force inspection system is part of a broader governance structure.[1] And the frequency of external inspection events is one of many moving parts in the design of the Air Force inspection system. To choose the best inspection interval, SAF/IG needs to understand (1) the role that the inspection system plays in the Air Force's broader governance system and (2) the role that the inspection interval plays in the Air Force's inspection system. In this chapter, we place the frequency of external inspection events within the broader governance context. We use information we collected from Air Force inspectors and inspectees to suggest what pattern of frequency they think would be most appropriate for the Air Force. We then contrast the Air Force and the FAA perspectives on inspection intervals. Inspection intervals in the FAA appear to be designed more effectively as part of a broader governance structure than those now under consideration in the Air Force.

Inspection Frequency and Governance

Every large, complex organization uses a variety of strategies to align its elements to a common purpose. One way of assessing whether any set of strategies is likely to be compatible with best practice is to build a simple model to predict what a cost-effective set of mechanisms would be in any particular setting. In this section, we apply this approach to the choice of inspection interval by placing it in a broadly used, standard economic model of organizational design and examining how the interval would vary depending on an organization's specific operating environment or priorities.

Using this approach, we can posit that any large, complex organization gives each of its components (for example, the wings of the Air Force)[2]

- *information* about how their activities can and do affect the well-being of the larger organization

---

[1] It is tempting to refer here to a "command and control system" rather than a "governance system." But "command and control" is a term of art within the Air Force and DoD that can mean very different and specific things to different people. To avoid slipping into the ongoing (useful and substantive) debate about what command and control means in DoD, we use the term "governance system," which organizational analysts use more generally to refer to any system that a large, complex organization uses to align all of its component parts to a common purpose. Its inspection system is surely one of the key mechanisms that the Air Force leadership uses to perform this responsibility. For a useful discussion, see Alberts and Hayes, 2006.

[2] This framework is based on Arrow, 1974, and Jensen, 1998. For recent practical applications of this framework in a variety of government settings, see Bartis, Camm, and Ortiz, 2010, and Stecher et al., 2010.

- *local capabilities*, like resources, training, local leadership, standard operating procedures, databases, models, and decision-support tools, that the component can use to refine the information it receives and link it to local information about its own activities
- *local motivation*, in terms of shared cultural values, and more or less formal incentive systems that reward those local components that actively promote the larger organization's performance and punish those that hinder the larger organization's performance.

Taken together, we can think of the specific mechanisms an organization uses to co-align its constituent parts—information, capabilities, and motivation—as a governance structure. The Air Force Inspection System is one element in such a structure. For example, the Air Force leadership uses its inspection system to give wings clear guidance—AFIs and checklists—about what it expects them to do to promote the performance of the Air Force as a whole. Inspections, assessments, and evaluations provide detailed information on the degree to which a wing is following that guidance and, implicitly, what elements of that guidance the leadership gives most priority. In terms of capabilities, leadership can give the wings training on self-inspection, access to best practices, data management tools like MICT, leaders and personnel with experience applying these tools, and so on to help wings enhance Air Force–wide performance. With respect to motivation, the Air Force leadership can, in principle, motivate performance *within a wing* by (1) emphasizing some definition of performance among the wing's priorities that is precisely tied to the guidance above and (2) potentially giving personnel in better-performing wings or squadrons better promotion opportunities.[3]

Frequency of external inspection events can play an important role in each part of such a governance structure in that

- greater frequency allows more frequent information collection and feedback about a wing's performance relative to the leadership's priorities
- greater frequency allows more opportunities to detect and diffuse information about best practices—and hence basic capabilities—across wings
- lower frequency allows a wing's focus on its primary military missions to be less often disrupted and, all else equal, allows the wing to improve its performance in ways that give members of the wing greater self-confidence and promotion opportunities, and earn the wing more attractive missions.

The Air Force leadership need not use its inspection system to accomplish any of these outcomes. *To the extent that it wants to*, however, the leadership's choice of frequency of external inspections will directly affect the governance structure it uses to command and control wings. As the factors we consider in this chapter become more important, the Air Force leadership will be more inclined to use its inspection system to stimulate each element of its broader governance system.[4]

---

[3]   The emphasis here is on the governance of the wing. We recognize that the Air Force views individuals not just in the context of their service at any one wing, but in terms of their development throughout their careers. The idea here is to give individuals more precise guidance on what is expected of them in any assignment and then judge their performance in that assignment against the guidance given.

[4]   If the Air Force does not view its inspection system in this light—for example, if inspection every two years is an end in itself—the analysis presented here does not apply.

Below are some patterns that should emerge in organizations that closely integrate the frequency of external inspection with their broader governance structures.[5]

External inspection becomes less frequent when the activities to be inspected demonstrate better local capability and/or motivation to promote the broader organization's goals in the absence of such inspection. In the Air Force setting, local leadership can be thought of as a local capability. Unit discipline can be thought of as a local capability in itself or as a shared set of values that motivates the alignment of local and broader organizational goals.

For analogous reasons, given the local capabilities and motivation of any organizational element subject to inspection, external inspections become more frequent as they are able to extract and provide useful information at a lower cost to the inspectors and the component being inspected.[6] More effective sampling could make external inspections more cost-effective for the Air Force and so encourage more frequent external inspections. More costly inspections, on the other hand, encourage increased inspection intervals.

In organizations with local components that have proven themselves capable and motivated, external inspections can become less frequent and more focused on processes that support this capability and motivation rather than details about outputs or resource consumption. This might mean focusing more on audits of a wing's local quality control system (a capability) or the state of unit discipline or leadership (motivation or capability).

Over time, broader organizations can make the frequency and focus of each component's inspections conditional on observed past performance of outputs, resource consumption, and processes. If performance remains high, the broader organization can extend the interval between inspections and focus more on local processes. On the other hand, if performance slips, the broader organization can tighten the interval of external inspections and make these inspections more aggressive and invasive.

Such an approach has two different but complementary effects within a broader governance structure. First, it uses external inspection as a safety net to avoid bad outcomes for the broader organization if local capabilities or motivation falter. Second, it rewards a local component's good performance and punishes the bad by scaling the degree of external intrusion to that performance. The first effect is relatively more important in the Air Force setting than it is in nonmilitary settings, since the frequent turnover of military personnel could mean that the interval between external inspections is conditioned on the past performance of an entirely different set of personnel. The disparity between the two effects grows as the inspection interval grows relative to the rate of turnover.

---

[5] These patterns are compatible with the predictions of a microeconomic model of the quality of the governance of one local organizational component part as the product of three types of inputs: information, local capability, and local motivation. Increasing any of these inputs while holding the other two constant increases the quality of governance. A given level of quality of governance can be achieved by a wide range of different combinations of information, local capability, and local motivation, which all act as substitutes for one another. The relationships among these parts of the governance structure display all the scale and substitution effects one would expect in a standard, well-behaved microeconomic production function. The discussion of these relationships in the text assumes that the desired quality of governance is driven primarily by performance (effectiveness) requirements, not affordability (efficiency) considerations. The demand for quality in a governance structure is driven primarily by the criticality of the activity subject to inspection; demand for quality rises with criticality. For a given level of criticality, the level of quality of governance demanded changes relatively little with changes in the cost of providing quality governance (this demand is relatively "price insensitive").

[6] They become a more cost-effective element of the broader governance structure and so can be cost-effectively substituted for local capabilities and motivation within any organizational component part subject to inspection.

Not all external inspections need to be conducted in the same way. For example, improved information systems can allow virtual external inspection on an ongoing basis with little direct intrusion. As one officer noted,

> The technology is there. SharePoint is there. You could set it up to send trigger [compliance-related] email reminders, like reminders to post meeting minutes every month or wash the commander's car. . . . Lots of electronic things can be done, especially with training. It can be looked at from far away. You don't need 30 inspectors on base to look through the ADLS [Advanced Distributed Learning Service]. Create an electronic binder, put it on a CD, and give it to the inspectors (Inspectee, officer focus group 1).[7]

We heard a similar perspective in another focus group:

> If you could upload virtual documentation that they could look at without having to come, that would be helpful. There could be a version of that that would help out units. There could be some benefit in cutting the amount they have to look at when they do come out (Inspectee, officer focus group 2).

Where such systems can monitor local performance reliably, the intervals between more intrusive face-to-face external inspections can be extended or, more likely, made contingent on performance observed from a distance. The focus of inspections can also be limited only to collecting information that cannot be obtained from a distance and to assessing those areas in which virtual observation indicated significant problems. This pairing of non-intrusive virtual inspection with contingent face-to-face inspection could make governance structures within the Air Force's high-turnover setting more effective by conditioning the use of more intrusive inspections on the performance of the military personnel in place at any particular time.

It has long been a standard Air Force tenet of structural integrity analysis that the faster a crack can propagate in a particular material or application, the more frequent scheduled inspections and maintenance become. Similarly, the maintenance of a governance structure becomes more crucial when local circumstances relevant to the broader organization can degrade more rapidly. More frequent inspection is one method of maintaining the quality of governance; greater attention to local capabilities and motivation can also presumably help.

The effects above hold for all activities subject to external inspection. The higher the criticality of the activity, however, the larger the effect its performance will have on the broader organization (by definition). To temper this effect, the broader organization seeks a capable and reliable governance structure. More frequent inspection is one change that can improve the quality of a governance structure. Better local capability and motivation can as well. We would expect to see even more attention given to each of these factors in the inspection of a nuclear activity than in that of a non-nuclear activity. Similarly, activities involving flight safety or ground personnel safety should be prioritized over the quality of groundskeeping or quality of life.

---

[7]   After each quotation, a unique identifier indicates the interview or focus group session in which the comment was made. The same identifier is used to denote the same session throughout the report, but it does not have significance nor can it be used to identify the interview or focus group. These numerical identifiers are used to convey the extent to which evidence is present in multiple sessions.

Administrative simplicity is valuable in any large, complex organization like the Air Force. Such simplicity can clarify priorities, particularly in the face of a dynamic, uncertain external environment, relatively rapid personnel turnover, and heavy reliance on a comparatively junior workforce. There is a natural limit to the Air Force's ability to fine-tune the frequencies of its inspections, assessments, and evaluations to take account of all these unique circumstances. Some compromises will be required when writing policy to ensure that inspection intervals are feasible and reliable. These compromises, however, can adjust over time if significant changes in circumstances occur.

The frequency of inspection is one element in a governance structure with many moving parts. Different governance structures make sense in different settings. As a result, the appropriate inspection interval for a specific activity depends on its criticality, its observed recent performance, the cost of external inspections, the quality of other parts of the governance structure like local capabilities and motivation, the effect the cost of inspections has on local motivation, etc. It is unlikely that any one interval is appropriate for all circumstances, especially considering that these circumstances can easily change over time.

How Air Force Personnel in the Field View the Inspection Interval[8]

We heard a wide range of views from Air Force personnel on what the most appropriate inspection interval would be. Many noted that, in current practice, assessment and inspection intervals vary dramatically across different MAJCOMs and types of activities in the Air Force. Our participants' differences of opinion often explicitly reflected their different circumstances and what they thought was important in these particular circumstances.

The most common response from Air Force personnel was that a Unit Compliance Inspection (UCI) should occur every 24 months to align the inspection interval with the standard tour length of wing leaders. We frequently heard that "every commander should face a major inspection at some point in time." When exactly during a commander's tour should such inspections occur? On this question, there was no consensus. Some felt the inspection should be scheduled shortly after a commander arrived to inform him or her of how well the unit was performing and set a benchmark or baseline for the remainder of his or her tour. Others felt the inspection should come near the end of a tour, when it could best measure the commander's achievement on that tour, thereby motivating him or her to build the unit's capability to look as good as possible during the inspection. A few pointed to what would be, in effect, a compromise. They supported holding the inspection about halfway through a tour to motivate the commander to learn the new job quickly and give him or her a performance baseline to work from in the second half of the tour.

Still others suggested that the timing within a tour was less important than simply ensuring that some major inspection event occurs during the tour. If anything, these personnel thought timing should depend on the mission, deployment responsibilities, and commitments of the unit to avoid distracting attention from the mission at critical times. Unlike many others we talked to, these personnel appeared to serve in units that did not have some portion of their capabilities deployed at all times.

---

[8]   All normative or prescriptive statements in this section reflect statements we heard from Air Force personnel, not RAND's independent judgment.

Commitment to a 24-month interval was the dominant response, but many thought priorities other than timing could determine the inspection interval. In general, these people believed that some major inspection should occur at least every 18 months, alternating between different types of inspections. So, for example, a compliance inspection might occur every 36 months and a readiness inspection might occur every 36 months, but one or the other should occur at least every 18 months. We heard considerable support for such an approach if separate compliance and readiness inspections were to persist. Here is how one airman put it:

> With a three-year cycle, if you have two types of major inspections and time them 18 months apart, then everyone will be hit with at least one major inspection. It will not be a cloud over the commander, but there will always be one [inspection] on the calendar. Knowing you are vulnerable to being inspected [is important]. You never know where there will be a few bad commanders, and you don't want them to escape because they got through without an inspection (Inspectee, leadership interview 10).

Some participants supported such a pattern, but with a shorter interval between inspections. When we probed, we heard three different justifications for this approach, but no one offered solid empirical evidence to support these arguments. In fairness, we did not probe our participants for evidence. Some pointed to their own recent experience with alternative inspection regimes and drew conclusions based on their judgment of the effects of these regimes. It is probably best to regard their views as professional military judgment based on recent personal experience.

First, personnel were concerned that effective compliance would fall in the absence of frequent inspections. In the simplest terms, one focus group participant stated, "We are all human; we want to slack off when it [the inspection] is over. You want to take a month off, which becomes two months, which becomes a year" (Inspectee, NCO focus group 5). Another broadly held belief was that "[i]f you . . . don't review them [a unit] . . . for three years, for the next two years and 11 months, they will lose focus until they have to ramp up again. It's easier for them to always have the notion of an upcoming inspection, so they don't get lazy" (Inspector, focus group 2). Put more bluntly, when consolidation of inspections lengthens the inspection interval,

> An obvious benefit is more white space. An obvious consequence is more white space— more time for things to go wrong. So if you combine to get rid of inspections so that you just do one UCI every four years, then I guarantee that everything will be wrong at two years (Inspectee, office focus group 2).

Second, some people promoted less detailed inspections that would reduce the burden on inspected units and inspection costs. But, they cautioned, it would also probably reduce the degree of compliance. To compensate for this effect, these people recommended more frequent inspections. On net, they expected more frequent, less detailed inspections to yield better compliance over time relative to inspection cost, or vice versa.

Third, some people were concerned that inspection checklists should be updated more frequently to reflect the changing environment within which a unit operates. Various people noted that external regulations, weapon systems, and technology in general are changing more rapidly today than they have in the past. As a result, checklists should be updated more frequently than they have been in the past. And, if checklists are being changed more frequently, inspections should occur more frequently to ensure that all units are in compliance with the most recent version of each relevant checklist. One person summed it up this way:

> Technology changes all the time, and systems are constantly reviewed and sped up. Every four years would have been okay back in the WWII [World War II] or Vietnam era when we were still using typewriters and carbon copies. Now we have computer-based learning, databases, etc. . . . . When technology changes, the requirements change. We need to inspect more often because technology changes. And the CCOs [Core Compliance Objectives] change (Inspectee, NCO focus group 3).

In contrast, other participants gave us cogent arguments for why inspections should occur less often. Three distinct lines of thought captured most of the arguments we heard for longer intervals between inspections. First, intervals should be lengthened as the costs of inspections—both to the units and to the inspectors and assessors—increase. The costs to a unit include negative effects on the unit's mission. For example, if the unit's current mission is more important than an alternative mission being examined by readiness inspections or than its compliance goals, a longer interval could be justified. As one person said, "We don't exist to be inspected; the mission is why we exist. Inspections can help us with the mission, but not so frequently" (Inspectee, leadership interview 10).

Second, the more effective a unit's self-inspection system, the longer intervals can be between external inspections. In effect, this argument suggested that compliance was less likely to fall during longer external inspection intervals if a self-inspection system maintained the accountability of the unit's leadership or at least improved the leadership's knowledge of the unit's compliance.

Third, intervals between external inspections could be extended if performance on previous inspections had been exemplary. Participants tended to present this as a reward for good performance and hence an incentive to promote such performance in the future. One said, for example,

> I think it should be set like PT [physical training]. If you get an "excellent" or higher, you get fewer inspections. If you get a "marginal" you get re-inspected sooner. Provide a benefit to being "excellent" (Inspectee, leadership interview 11).

Again, participants did not offer evidence to support their assertions.

Scheduling Inspections at the FAA

To provide a context for the Air Force personnel's opinions of appropriate inspection intervals discussed above, we contrast their perspectives with the approach used by the FAA, a federal agency for which inspection is a core competency. The FAA's approach to scheduling inspections is fundamentally different from any of the processes discussed previously in this chapter. Aircraft operators ("air carriers"), repair stations, designers, and manufacturers must hold certificates from the FAA to operate in the United States. The FAA uses a two-phase process to certify members of the aviation community. The Design Assessment (DA) phase of this process ensures that carriers' operating aviation systems comply with FAA regulations and safety standards. The Performance Assessment (PA) phase confirms that carriers' operating systems produce their intended results. In this section, we describe these two types of inspections as well as how the FAA sets their schedules.[9] We then relate them to the Air Force inspection system.

The Design Assessment Phase and Safety Attribute Inspections (SAIs)[10]

As noted above, the DA phase ensures that carriers' operating systems comply with FAA regulations and safety standards, including the requirement to provide service at the highest level of safety in the public interest. A poorly designed system compromises safety and the ability to perform safety risk management. DA is the most critical phase of the FAA certification process because it ensures that each carrier has a properly designed operating system. SAIs evaluate the design quality of carriers' operating systems in terms of six safety attributes:

1. procedures—documented methods to accomplish a process
2. controls—checks and restraints designed to ensure a process's desired result
3. process measures—metrics used to validate a process and identify problems or potential problems in order to correct them
4. interfaces—interactions between processes that must be managed to ensure desired outcomes
5. responsibility—a clearly identifiable, qualified, and knowledgeable person who is accountable for the quality of a process
6. authority—a clearly identifiable, qualified, and knowledgeable person who has the authority to establish and change a process.

The FAA's Principal Inspectors (PIs) collect data on these attributes to make informed judgments about the design of carriers' operating systems (1) before approving them according to FAA regulations and (2) during recurring assessments of continued operational safety.

In general, the FAA conducts SAI inspections of a carrier's operating system design every five years, unless (1) significant deficiencies are found during more frequent Element Performance Inspections (EPIs) (see below), (2) significant changes in a carrier's management personnel or organizational structure have taken place, or (3) there are significant changes in carrier regulations.

The Performance Assessment Phase and EPIs

The PA phase confirms that carriers' operating systems produce intended results, including the mitigation of risks associated with significant hazards. It assesses whether carriers follow the written procedures and controls documented in the DA phase and meet FAA's established performance measures for all operating systems. EPIs analyze how carriers' operating systems,

---

[9]   The FAA uses different processes to conduct this risk assessment for different kinds of certificate holders. The approach described here applies to scheduled air carriers ("Part 121 carriers"), which account for most U.S. air passenger and cargo traffic. FAA Order 8900.1A defines this approach. A similar risk-based approach is available for non-scheduled charter carriers (Part 135) and repair stations (Part 145). For aircraft in the design and production phases, FAA Order 8120.2G establishes "Risk Based Resource Targeting (RBRT)," a risk-based approach similar to the one for Part 121 carriers. All of these processes draw on a common Risk Management Process that applies a traditional probabilistic risk assessment (PRA) methodology. Appendix C describes this process.

[10]   The material in this section is based on official FAA documents and interviews with knowledgeable officials in the FAA. A particularly useful one-stop source of information is FAA, 2007.

subsystems, and elements interact. The FAA structures EPIs around seven air carrier operating systems:

1. aircraft configuration control—systems used to maintain the physical condition of the aircraft and its components
2. manuals—systems, such as information and instructions, used to define and govern the air carrier activities
3. flight operations—systems pertaining to the movement of aircraft in flight
4. personnel training and qualifications—systems used to ensure proper personnel training and qualifications
5. route structures—systems used to maintain facilities on approved routes
6. airman and crewmember flight, rest, and duty time—systems that prescribe time limitations for employees
7. technical administration—systems used to address other aspects of certification and operation, such as key management personnel.

Each of these seven systems has a defined set of operating subsystems and operating elements within these subsystems (described in more detail below).

The FAA sets a separate EPI interval for each operational element within a certified air carrier. To determine the frequency of EPIs for a particular operational element, the FAA proceeds in three steps. (1) It starts with an assessment of the *inherent risk* associated with the element. (2) It then adjusts this inherent risk for *conditional risks* associated with the circumstances the FAA observes at each carrier. This adjustment provides the base "required" inspection interval, which the FAA can execute only if it has sufficient resources. Finally, (3) the FAA allocates its available inspection resources across all of the requirements identified in the second step to determine the actual inspection frequency for each operational element at each air carrier.

Step 1: Inherent Risk. The FAA maintains a set of pre-established inspection frequencies for different operating elements within carriers' operating systems and subsystems based on the inherent risks it associates with performing that operating element, regardless of which air carrier performs it. Table 2.1 presents a comprehensive list of operating systems, subsystems, and elements for Part 121 carrier certifications pertaining to airworthiness. Each operating element is rated as having a high (red), medium (green), or low (blue) risk factor based on its associated inherent risks. Table 2.1 shows, for example, that several elements related to airworthiness certification in the maintenance organization (e.g., required inspection items, maintenance providers, major repairs and alternations, reliability programs) are inherently risky, requiring more attention than elements in other subsystems.

To establish this rating structure of inherent risks, the FAA engaged subject matter experts (SMEs) in the mid-1990s to review air carrier operating systems, subsystems, and elements.[11] The FAA has continuously refined and modified these ratings since.

The FAA's current guidelines require that elements with a high criticality risk rating (red) be inspected every six months. Those with medium (green) and low (blue) risk ratings are required to be inspected every 12 months and three years, respectively. Again, this interval

---

[11] This was one response to the ValuJet crash in 1996. Chapter Seven discusses the role this crash played in the FAA's broader program to ensure aviation safety.

Table 2.1
Operational Systems, Subsystems, and Elements Relevant to Air Worthiness

1.0 Aircraft Configuration Control	
1.1 Aircraft	
1.1.3	Special Flight Permits
1.2 Records and Reporting Systems	
1.2.1	Airworthiness Release/Maintenance Log Recording Requirements
1.2.4	Mechanical Interruption Summary (MIS)/Service Difficulty Report (SDR)
1.3 Maintenance Organization	
1.3.1	Maintenance Program
1.3.2	Maintenance/Inspection Schedule
1.3.3	Maintenance Facility/Main Maintenance Base
1.3.4	Required Inspection Items (RII)
1.3.5	Minimum Equipment List (MEL)/Configuration Deviation List (CDL)/Deferred Maintenance
1.3.6	Airworthiness Directives and Maintenance Record Requirements
1.3.7	Maintenance Providers
1.3.8	Control of Calibrated Tools and Test Equipment
1.3.9	Major Repairs and Alterations
1.3.10	Aircraft Parts/Material Control
1.3.11	Continuous Analysis and Surveillance System (CASS)
1.3.15	Reliability Program
1.3.16	Fueling
1.3.17	Weight and Balance Program
1.3.18	Deicing Program
1.3.19	Lower Landing Minimums
1.3.23	Short-Term Escalations
1.3.24	Coordinating Agencies for Supplier's Evaluation (C.A.S.E.)
1.3.25	Cargo Handling Equipment, Systems and Appliances
2.0 Manuals	
2.1 Manual Management	
2.1.1	Manual Management
4.0 Personnel Training and Qualifications	
4.1 Maintenance Personnel Qualifications	
4.1.1	RII Personnel
4.1.2	Maintenance Certificate Requirements

Table 2.1—Continued

4.2 Training Program

4.2.1	Maintenance/Required Inspection Item (RII) Training Program

5.0 Route Structures	

5.1 Approved Routes and Areas

5.1.1	Line Stations
5.1.8	Extended Operations (ETOPS)
5.1.9	Reduced Vertical Separation Minimum (RSVM) Authorization

7.0 Technical Administration	

7.1 Key Personnel

7.1.1	Part 119 Required Personnel
7.1.6	Maintenance Control

NOTE: Red element = high criticality, green element = medium criticality, blue element = low criticality.
SOURCE: FAA, 2007, Figure 10-6.

is a *starting point* to determining the inspection interval the FAA will actually apply to each operational element.

Step 2: Conditional Risk. FAA PIs assess each carrier's actual operating systems and their operating environment for indications of safety hazards or conditions that may create risks unique to the carrier. To do this, the PIs review the 28 risk indicators contained in the Air Carrier Assessment Tool (ACAT) and shown in Table 2.2. The FAA categorizes these indicators in terms of

- environmental criticality (EC)—aspects of the air carrier's surroundings that may lead to or trigger a systemic failure with the potential to create unsafe conditions
- performance history (PH)—results of the air carrier's operations over time
- operational stability (OS)—organizational and environmental factors the air carrier cannot directly control, but can manage effectively to improve system stability and safety
- air carrier dynamics (CD)—organizational and environmental factors that the air carrier can directly control to improve system stability and safety.

The PIs determine a risk score (1 through 6) for each risk indicator in Table 2.2 for each air carrier. SMEs have developed detailed explanations of how to assign this score.[12] The FAA has continuously refined and modified these explanations since their initial development in the mid-1990s.

Table 2.3 summarizes the scoring explanations for "outsourcing risk," one of the environmental criticality risk factors. The commercial aviation industry is increasingly outsourcing traditional air carrier functions to independent contractors, which the FAA believes increases

---

[12] Note that risk-scoring schemes for each risk indicator are predefined but PIs assign risk scores for each risk indicator area for a given carrier whenever a new certification cycle starts.

potential risks. Outsourcing has developed to the point where multiple levels of contractors could be involved in providing a single service, adding still more risk.

Step 3: Risk-Based Inspection Interval. The ACAT combines (1) the inherent risk critical- ity ratings of each operating element described in the previous section with (2) the PI's assigned risk score for each risk indicator area. It then automatically provides an inspection priority

Table 2.2
Air Carrier Assessment Tool (ACAT) Risk Indicators

	ENVIRONMENTAL CRITICALITY (EC)	
EC-01	Age of Fleet	The age of the fleet can impact carrier's operating systems.
EC-02	Varied Fleet Mix/ Configuration	A varied fleet mix and/or mixed fleet configuration can significantly alter carrier's safety profile and the potential for failure in its operating systems.
EC-03	Change in Aircraft Complexity	Changes to the complexity of carrier's fleet can significantly affect carrier's safety and the potential for failure in its systems.
EC-04	Outsource (Maintenance, Training, Ground Handling)	The use of outsourcing programs, depending on a number of factors, could heighten the risks associated with various carrier operations. These programs must be effectively managed.
EC-05	Seasonal Operations	Short-term operations may present their own unique risks and may require attention and preparation by the carrier.
EC-06	Relocation/Closing of Facilities	Relocating or closing a facility may adversely affect operational and system stability of the carrier.
EC-07	Lease Arrangements	Aspects of lease arrangements may be sources of risk at the carrier and must be effectively managed.
EC-08	Off-Hours Activity	Carrier management of off-hours (i.e., as outside normal FAA hours, including weekends) activity can be prone to risk.
	PERFORMANCE HISTORY (PH)	
PH-01	Enforcement Actions	Enforcement actions can help identify carrier's safety profile and any area of risk in its systems.
PH-02	Accidents/Incidents/ Occurrences	Data regarding accidents, incidents, and occurrences may provide insights into areas of risks at the carrier.
PH-03	Department of Defense (DoD) Audits	DoD audit findings help to identify hazards and their associated risks. Audit data may provide insights into systemic problems in the design and performance of the carrier's systems.
PH-04	Voluntary Disclosures	The type and content of carrier's self-disclosures, and the effectiveness of the carrier's corrective actions can assist in the risk assessment.
PH-05	Safety Hotlines/Complaints	Excessive or repetitive safety hotline and other complaints against the carrier may assist in identifying and isolating areas of risk. Complaints can aid the carrier in managing and controlling corrective and follow-up actions.
PH-06	Voluntary Programs Data	Carrier voluntary program data may be useful for hazard or risk identification. Such data can aid the carrier in managing corrective and follow-up actions.
PH-07	Surveillance Indicators	Surveillance data from the Safety Performance Analysis System (SPAS) Program Tracking and Reporting Subsystem (PTRS), and the Air Transportation Oversight System (ATOS) help to identify trends in carrier performance and can assist with identifying risks in carrier's system design.

Table 2.2—Continued

		OPERATIONAL STABILITY (OS)
OS-01	Key Management SPAS Indicators	Changes in key management personnel can significantly impact carrier's system and operational stability.
OS-02	Financial Conditions	Carriers that experience adverse financial conditions may have higher risk.
OS-03	Change in Air Carrier Management	Changes in management personnel other than key management can significantly impact carrier's system and operational stability.
OS-04	Turnover in Personnel	A high turnover of operations and maintenance personnel can dramatically increase the potential for risk in carrier's systems.
OS-05	Reduction in Work Force	A reduction in carrier's work force can dramatically increase the potential for failure in the carrier's systems.
OS-06	Rapid Growth/Downsizing	Times of significant change such as rapid expansion or downsizing can impact carrier operations due to the possible misalignment of resources and operational requirements.
OS-07	Merger or Takeover	Carrier must effectively manage mergers or takeovers to ensure continued compliance and safe operating practices.
OS-08	Labor-Management Relations	A poor or deteriorating labor-management relationship can create risk.
		AIR CARRIER DYNAMICS (CD)
CD-01	New/Major Changes to Program	Safety issues may develop from new or changed programs and may increase the potential for noncompliance with existing processes and controls.
CD-02	Continuing Analysis and Surveillance System (CASS; AW Only)	Carriers with a poorly functioning Continuing Analysis and Surveillance System (CASS) can overlook and improperly manage increased levels of risk.
CD-03	Safety Management	Carriers who do not have a safety management system may not understand or adequately control hazards to operational safety.
CD-04	Relationship with the FAA	Carrier's relationship with its assigned FAA personnel may provide insights into the carrier's compliance posture and safety culture.
CD-05	Human Factors	Risk may exist due to human lapses in the carrier's design and/or performance.

SOURCE: FAA, 2007, Figure 10-16.

ranking of the operating elements based on the combined risk assessment. Based on this priority ranking, the PIs set the inspection schedule for each of the operating elements and identify "required" (i.e., unconstrained) inspector resources for each of the scheduled inspections. FAA managers then compare the required inspection resources to those available and, based on the priority ranking, determine the final resource allocations.

Implications for the Air Force

It is important to remember, when comparing the FAA and Air Force inspection systems, that they monitor different things. The FAA inspection system focuses on safety; the Air Force inspection system looks at safety as well, but can also look at anything else mentioned in AFI. Where the FAA assesses the consequences and likelihood of bad safety outcomes, the Air Force assesses the consequences and likelihood of bad outcomes for a broader array of performance attributes, safety among them.

Table 2.3
Outsource Risk-Scoring Scheme Under Environmental Criticality (EC-04)

Risk Score	Inspector Considerations
1–2	• The carrier does not outsource. • The carrier's oversight staffing and audit functions appear to be adequate to include the outsourced functions. • The contractors effectively meet the training requirements of the carrier and appear to be qualified for the outsource function(s). • The carrier effectively manages impacts.
3–4	Concerns exist regarding the impact of outsourcing due to considerations such as: • the contract personnel are utilized by numerous air carriers increasing the possibility of non-adherence to procedures • the contract personnel training records are inaccurate • adverse DoD findings against the contractor • contractor qualifications and abilities (maintenance, training, and/or ground handling) are in question • the carrier frequently changes contractors based on economics and/or • the use of outsourcing for all or particular functions is relatively new at the carrier; therefore, lack of historical data is a major consideration.
6–7	• Concerns exist about the impact of outsourcing because the air carrier does not have an effective safety audit function to monitor the performance of the contractors. • Concerns exist because the contractor's performance history indicates multiple, repeated safety violations.

SOURCE: FAA, 2007, Figure 10-20.

Setting that difference aside, the FAA approach to scheduling inspections differs from both the current and proposed Air Force methods in two fundamental ways:

- FAA inspectors use many different inspection intervals in their oversight of different elements of the aviation system.
- FAA inspectors use different inspection intervals in their oversight of different certificate holders, even when examining exactly the same elements of the aviation system in different places.

Some portions of the commercial aviation system are simply less risky than others. They pose a smaller potential for contributing to large negative outcomes in the aviation system, or are less likely to have significant negative impact of any kind. The result is that the FAA is regularly prepared to inspect some activities five times more often than others. An FAA inspection team may look at one certificate holder every six months, using a different set of checklists each time, to assess how well the certificate holder is actually executing the operational system it has designed.

The principal factor currently considered by the Air Force when choosing the inspection interval for a new system is the normal tenure of management across its aviation system. In determining its inspection intervals, the Air Force gives almost no direct attention to how the risks associated with each element of its aviation system differ. Even though Air Force inspectors clearly think about the relative criticality of different elements when making sampling decisions during inspections (see Chapter Three), they have no guidance as clear as the FAA's for determining how critical various operational elements are relative to one another.

Unlike the Air Force, the FAA routinely adjusts all inspection intervals to reflect local circumstances of individual certificate holders. If the EPIs that monitor conditional risk at relatively short intervals indicate that a certificate holder is no longer executing its process as designed, the FAA can shorten the normally long—five-year—inspection interval of the SAIs that monitor inherent risk in the design of operational systems. If EPIs indicate that a certificate holder's performance has fallen below FAA standards in particular places, the FAA can shorten the EPI intervals of only those elements with degraded performance. That is, inspections can be enhanced and resources applied specifically to those areas where experience suggests increased value. If industry-wide revenue shortfalls tempt certificate holders to spend less on maintenance, the FAA can proactively shorten the interval between EPIs relevant to maintenance to forestall accidents and other serious incidents.

The Air Force does not adjust its inspection intervals to reflect any factors of this kind, nor did it plan to at the time of this report's completion (2011). Perhaps because all Air Force inspections and assessments use common checklists, it is hard to enhance the monitoring of only specific items. The difficulty of targeting emphasis makes it less cost-effective for the Air Force to vary its inspection emphasis at all over time or between wings. Again, individual inspectors do this where they have discretion, but they cannot determine inspection intervals.

Additionally, if budget shortfalls tempt the Air Force to spend less on maintenance, it will probably decide to cut resources for inspections and assessments as well—maybe even more than those for maintenance. The Air Force inspection system has no natural way of adjusting its inspection intervals to provide pushback to changes in aviation system resource that could degrade performance in the Air Force as a whole.

The FAA also adjusts inspection intervals when changes in its own policies destabilize the environment in which certificate holders operate. By contrast, the Air Force inspection system does not shorten inspection intervals in areas where its policies have changed, or even where its weapon systems or operating policies have changed.

In sum, the FAA inspection system exhibits much more variation than that of the Air Force. In effect, the Air Force places a higher premium on simplicity than the FAA does. This is a bit surprising because, as a regulatory agency, the FAA must give special attention to the dictates of administrative law. These dictates tend to encourage simplicity as a means of avoiding protests based on the unfair treatment of regulated entities, or the lack of due process or transparency in administrative procedure. Such legal constraints do not influence how the Air Force inspection system treats different wings and different activities within wings. Perhaps broad norms about fair treatment within the Air Force trump this difference. Or perhaps Air Force personnel are so much less experienced than the commercial aviation workforce that they require greater simplicity.

Summary

The inspection interval is just one element of the Air Force inspection system, which is itself an element of the governance structure the Air Force uses to align all its organizational components to work toward a common purpose. To integrate the Air Force's choice of inspection intervals into its broader governance structure in a cost-effective way, we can predict how a number of key factors—for example, the relative criticality of a mission, the relative capability of a particular wing, and the degree of stability in a wing's operating environment—would affect the length of the inspection interval in different circumstances.

We can make analogous predictions about how the FAA chooses inspection intervals in its oversight of the commercial aviation system. The FAA's choices are broadly consistent with what we would expect a cost-effective design for an inspection system to be—shorter inspection intervals for inherently critical or risky activities and for organizations in less stable operating environments, and longer intervals for organizations that have demonstrated better local capabilities. Many Air Force personnel we spoke to recommended a similar approach for the Air Force. The FAA inspection system can apply significantly different inspection intervals—over the span of an order of magnitude or more—to inspections of very specific items in a commercial company that display distinct characteristics relevant to governance.

Factors that affect cost-effectiveness shape the perceptions many Air Force personnel in the field have of inspection intervals. Changes to the Air Force inspection system currently under consideration, however, do not take these factors into account. The primary factor receiving attention in the design of the future Air Force inspection system is the nominal tour length of the leaders of a wing. The FAA considers such a factor in its deliberations, but only as one factor among many others that affect the inspection interval. As we shall see below, opportunities exist to reflect some of these other factors in the design of any inspection that occurs, even if the inspection interval is fixed.

CHAPTER THREE

Reducing the Inspection Footprint

One of the most important principles of the Air Force inspection system is to

> minimize the inspection footprint to the maximum extent practical [sic]. When deemed appropriate, MAJCOM/IGs may give credit for unit activity in conjunction with exercises and contingencies, real-world operations, self-inspections, combined inspections, other inspections/evaluations, sampling techniques, and other measures of sustained performance to minimize footprints. These measures can be used to adjust the breadth, duration, and incidence of inspection activities.[1]

We can think about "footprint" from two qualitatively different perspectives. One is the total number of manpower hours during a year that a wing spends preparing for and participating in external inspection events. More hours means a larger footprint. The other is the number of days during a year that a wing is subject to any external inspection event. The fewer days of inspection, the more "white space" a wing has to focus on its military mission, and the smaller the footprint. Current efforts to improve the Air Force inspection system tend to emphasize the second—"white space"—relative to the first—hours of time consumed—but both are important.

Many of the Air Force personnel we talked to believe that scheduled inspections divert a wing from its primary military mission. There are always more tasks for a wing to do than it has the resources to carry out safely. On a day-to-day basis, one of the primary tasks of a wing's leadership is to decide in which areas to take risks.[2] If the leadership emphasizes the wing's mission, risk will fall on those elements of compliance that do not relate directly to the mission in the form of less-stringent monitoring. If, on the other hand, the leadership emphasizes preparation for an external inspection event, some risk must be taken with the current military mission. That is true whether the inspection event emphasizes compliance risk issues not immediately relevant to the current mission or military missions other than the current mission that the wing must remain ready to execute as needed.

One might argue that improvement in the leadership of a wing can potentially yield better outcomes—whether measured in terms of the current military mission, alternative missions,

---

[1] AFI 90-201, 2009, p. 9, Sec. 2.1.2.

[2] This does not mean that wing activities driven by inspections always divert a wing from its primary mission. For example, ensuring that all personnel in the wing are receiving the training called for in an inspection checklist also presumably helps ensure that these personnel are qualified to execute the wing's primary mission. But when resources are scarce—as they typically are and as they certainly will be in the future—choices sometimes have to be made. At some point, ensuring that every administrative document is properly completed and up to date can consume resources that the wing might have used to increase operational readiness. The personnel we talked to argue that such conflicts often arise and must be resolved.

or other compliance issues—from any given set of resources. This is undoubtedly true. But the better a wing is run, the more risk the wing must take on one front—for example, a military mission other than the primary mission being executed—to reduce risk to another—for example, compliance. The Air Force personnel we talked to find themselves currently making this compromise. A heavy deployment schedule even further intensifies this sense of pressure.

External inspections also consume the time and resources of personnel who work full-time as formal inspectors, assessors, or evaluators as well as those who augment the full-timers. Inspectors we talked to find themselves in a situation similar to that of the wings they oversee. They strongly support efforts to improve the inspection system by reducing these costs because (1) they tend to believe that current Air Force inspection, assessment, and evaluation activities are already under-resourced and, (2) as the Air Force faces additional budget pressures in the near future, they believe they will likely have to give up additional inspection-related resources.

This chapter discusses four ways of reducing the inspection footprint:

- synchronize inspection events
- integrate inspection events
- use sampling to reduce the time spent assessing each activity during an inspection event
- use a higher percentage of no-notice inspection events.

By reducing the inspection footprint, we mean that at least one of the following three changes occurs: (1) the amount of time wings spend preparing for and enduring external inspection events is reduced; (2) the amount of white space that a wing has increases; and/or (3) the resources that inspectors consume when they prepare and execute events is reduced.

Table 3.1 summarizes the effects that such changes are likely to have. Each row addresses a different change. The columns name each change, and then show its potential positive and negative effects, as identified by the Air Force personnel we spoke to.

Efforts to improve the inspection system can potentially execute elements of each of these types of changes. The text in Table 3.1 can be thought of as (1) describing the individual effects of each change, or (2) describing the effects of applying all changes cumulatively, starting from the top of the table. Either way, it is important to understand that each change shapes the effects of the others. Any effort to pursue such changes should think about them in an integrated way. The remainder of this chapter discusses each of these changes and the relationships among them.

Synchronize Inspection Events

An obvious first step toward clearing white space is to schedule inspections and assessments to occur on the same day. Here is the perspective of one person eagerly awaiting synchronized arrangements:[3]

> Our UCI is books, and the LCAP [Logistics Compliance Assessment Program] is performance, but they go together. Books and task eval[uation] go together. We asked if we could get LCAP at same time as UCI, and they said no. Those inspectors are down the hall from

---

[3]   All normative or prescriptive statements in this section reflect statements we heard from Air Force personnel.

Table 3.1
Potential Effects of Efforts to Reduce the Inspection System Footprint

Option	Pros	Cons
Synchronize inspections	• Increases white space by scheduling together inspection events overseen by different Air Force entities.	• Increases logistical burden on base during any inspection event. • Increases complexity of scheduling.
Integrate inspections	• Reduces inspector and inspectee time by asking any one question or imposing any one task only once. • Potentially increases white space by reducing the length of any inspection. These effects increase as more synchronization occurs.	• Increases preparation time to allocate tasks. • Time to prepare increases as more synchronization occurs.
Increase use of deliberate sampling	• Increases white space by reducing the length of inspection events. • Reduces inspector and inspectee time consumed during an event.	• Reduces quality of inspection event by reducing the quantity or quality of information collected. • Increases preparation time for inspectors. • Cost-effective sampling balances preparation time with quality of information collected.
Use more no-notice inspections	• Smooths the path of inspectee preparation over time. • Increases compliance between inspection events. • Potentially increases white space if it allows less frequent inspection events. • Potentially decreases inspectee preparation effort over time.	• Reduces the level of performance that can be expected of inspectees. • Despite this, potentially increases inspectee preparation effort over time.*

*Some personnel expect no-notice inspections to reduce the burden of inspections on inspectees and others expect the opposite. Without more detailed information, we cannot know for sure why this apparent conflict exists, but their comments tend to be consistent with the following reasoning: The burden could *fall* if compliance falls off dramatically between scheduled inspections, requiring major efforts to prepare for scheduled inspections when they occur *and* it is relatively low cost to sustain compliance at an acceptable level if that level does not change much over time. The burden could *rise* if it is relatively costly to maintain compliance at an acceptable level all the time *and* it is not very hard to bring a noncompliant wing back into compliance in preparation for a scheduled inspection. If such reasoning is correct, one point of view might be correct in one setting and the other in another setting.

one another at ACC, but they don't talk. It would benefit us to have one massive inspection rather than two separate inspections. Now we have to prep for our LCAP somewhere around October, instead of getting back to the mission (Inspectee, NCO focus group 4).

Another person told us about external oversight events scheduled close to one another. This proximity presumably allowed the units being examined to prepare for the inspections and assessments at the same time. But this person wanted still more synchronization between the IG-led compliance inspection and smaller-scale, function-specific assessments: "Two weeks after the CI, we had the safety evaluation, ATSEP [Air Traffic System Evaluation Program], and SEPWO [weather inspection]. How can Air Traffic go through a compliance inspection only then to go through an Air Traffic inspection?" (Inspectee, leadership interview 1).

Many parts of the Air Force have been pursuing this synchronization strategy for some time. One inspector told us

[W]e're already there for Maintenance and LCAP. We are at the base at the same time as the other inspections teams 99 percent of the time. We essentially have overlapping or near-simultaneous inspections. We have to deconflict schedules [with other teams] as soon as we get there (Inspector, focus group 1).

As a result, many of the people we talked to had experience with at least one event of this kind. Experiences varied because the design of these events varies across the Air Force.

When events are synchronized, separate teams of inspectors and assessors typically arrive at a location at the same time or in close temporal proximity to one another, but conduct their affairs more or less separately. They typically have to coordinate their meeting schedules to avoid tripping over one another, share little information, communicate separately with the units being overseen, and finally issue separate, independent findings on the status of the units in question. In general, those we spoke with during interviews and focus groups have not been pleased with their initial encounters with such synchronization.

Such simultaneous visits can impose significant logistical burdens on the units being examined, especially smaller units. Smaller units often reside in the Air National Guard or Reserve. As one of our focus group participants told us, "Not every base can accommodate [such visits]. I was told to expect 120 IG inspectors, plus 35–40 LCAP inspectors. You can easily get 200 people on base at the same time, all needing vehicles, lodging, food. It can overwhelm the base and outstrip resources" (Inspectee, officer focus group 1). In particular, the unit logistical personnel responsible for hosting inspectors and assessors find themselves with a new mission at the same time that they are subject to external oversight. Another participant said, "It could be a logistical nightmare. Half of the Guard bases are small, and if you dump 50–100 people on a Guard base, you could get a 1:1 inspector/inspected ratio" (Inspector, focus group 1). Even someone from a moderate-sized active component base told us that

> Logistics would play a role. The logistics of housing and bringing more than 100 at a time onto [my base] and getting their rental cars and making sure they have somewhere to stay. We don't have the lodging. And the manpower to care for and feed them. And if they all require an operating station . . . (Inspectee, officer focus group 2).

These challenges can get worse when a significant portion of a wing's personnel is deployed:

> At any time, we're about 12- to 15-percent deployed. Getting prepared for this was painful with our deployment tempo. If you combine inspections, now you are increasing the questions that are covered. That could potentially break a unit depending on how you combine the questions. If you expand the inspection process, that could be too much (Inspectee, officer focus group 2).

Many suggested that synchronized inspection events would not be feasible at many bases without some integration. Experience to date has revealed very little integration of synchronized events. One person described the recent synchronization of a UCI and LCAP assessment that lacked such integration:

> The compliance inspection looks at programs and administrative processes. The LCAP are task eval[uation]s. Why are those separate? Why would a maintenance group get an inspection that looks at programs and procedures? Why would that be separate from a compliance inspection that looks at performance? I get that they are chartered under different organizations. [But] it makes no sense to me to have two inspections teams at the base at the same time that show no interest in sharing grades. Programs, admin[istrative]

process, performance ought to be conducted by the same team at the same time. It boggles my mind that we would think they are separate processes and thus make a deliberate effort to keep the teams separate, so the IG is not tainted by LCAP findings (Inspectee, leadership interview 1).

Another person told of a similar experience:

There is a mail manager inspection and the UCI. They inspect the same program. Mail manager never tells the UCI inspectors their results. They were both here the first week of June and they did not talk to one another (Inspectee, NCO focus group 4).

Some participants questioned the need to move every assessment activity on a base into the same schedule, especially when they involve a small, specialized portion of a wing. For example, one person talked to us about the standard, annual assessment of a base's firefighting capability. He said, "for Fire, the whole base should not have to coordinate, because they do it [the assessment] every year. It is kind of like a semester at school: a couple of quizzes and a final exam" (Inspectee, officer focus group 3). This participant felt that the exercise had no effect on the rest of the wing. Therefore, synchronizing it with other external oversight likely does nothing to increase white space at the wing. The same could be said of many specialized activities that normally operate well below the visibility of the wing commander.

Taken together, these observations support synchronization when it contributes materially to increasing a wing's white space and when effective integration measures limit the logistical and operational burdens that come with holding more inspection and assessment activities at the same time. Synchronization benefits are much more significant when accompanied by effective integration.

Integrate Synchronized Inspection Events

The benefits of integration are realized when fewer inspectors are required on site during an inspection and/or personnel in the unit being inspected spend less time and effort providing information. But as valuable as integration appears to be, our interview and focus group participants warned us that it is hard to do right.[4]

At first glance, integration can seem simple. We heard repeatedly that inspections and assessments of the same activities typically rely heavily on the same checklists. This makes sense, because functional communities create the checklists that inspectors generally apply in their inspections. For example, the checklists are the same for a UCI and LCAP. Some task evaluations differ, but otherwise they are very close. One working group found "an 80-percent duplication of questions on inspection checklists in the supply area" (Inspector, focus group 1).

The perception existed, however, that inspectors and assessors can take very different approaches to the same checklist item. In practice, inspectors generally tend to give more attention to unit compliance with required maintenance standards and documentation. They also tend to look at higher-level issues that cut across individual parts of a squadron or wing. This focus tends to give special attention to record keeping. Functional assessors, on the other hand, generally give more attention to evidence that personnel actually possess the skills documented

---

[4]   All normative or prescriptive statements in this section reflect statements we heard from Air Force personnel.

in wing records. They also tend to give greater substantive attention to the potential for process improvements in the specific activities they assess. A leader in an organization responsible for wing communications put it this way:

> Inspections are broken up now between managerial inspections and operational inspections, and the two have very different perspectives that do not lend themselves all the time to consolidation and simultaneous execution. For example, the CCRI [Command Cyber Readiness Inspection] is 100-percent operationally oriented. Its point is to achieve the most efficient communication structure possible. The UCI doesn't care about network efficiency; it cares about bandwidth and access. Combining the UCI and CCRI would be a mistake. There are options for economy, but this is not a "wholesale merge action" by any means (Inspectee, leadership interview 12).

So, in principle, even if the checklist items are identical for the inspection and assessment of the same unit, the inspectors and assessors could request very different information to determine the degree of compliance with any particular item. At one extreme, if an inspector and assessor approach a particular item from completely different perspectives, both will have to be present to achieve the full intent of the inspection and the assessment. At the other, if the inspector and assessor want exactly the same information, it is feasible to collect the information only once and save a significant amount of time for all involved.

Therefore, as several people pointed out, the Air Force can only determine the real potential for integration by walking through its checklists item by item to determine which benefit from the different perspectives of an inspector and an assessor. Even if there is significant value in these different perspectives, the Air Force must ask whether they need to be applied by different people. In many cases, an assessor serves as an augmentee to an inspection team and sees the same people at a wing twice—once as an assessor and once as an inspector. When this occurs, it seems plausible that one person could apply the perspectives of both an inspector and an assessor in a single transaction with the inspectees.

Perhaps the simplest way to coordinate between two oversight activities is for one to cede primary authority to the other. Several participants raised this possibility with regard to medical services. One person explained that the accreditation process is far more critical than any compliance inspection:

> For us, as a hospital, we have to get accredited. To think that the CI will help us go beyond [the accreditation] doesn't make sense. You bomb a UCI, you can still see patients, but if you bomb the accreditation, you can't see patients. Not even on the same scale. We work with the line. That is where it affects my shop. But as far as the UCI checking radiology, they have to be inspected by [the] health inspector (Inspectee, officer focus group 3).

Another person from a medical group being inspected agreed, stating, "The UCI generally says that we have an AAAHC [Accreditation Association for Ambulatory Health Care assessment], so they leave us alone" (Inspectee, NCO focus group 4). In another instance, a UCI team simply "left the medical group alone, since they would have their HSI [Health Services Inspection] soon" (Inspectee, leadership interview 6). Medical services are not the only area in which this happens. As an inspector told us, "some inspections overlap—UCI and ESOHCAMP [Environmental, Safety, Occupational Health, Compliance Assessment, and Management Program], for example. When we went out to do the UCI, we were told not to inspect the environmental portion, because the ESOHCAMP team was going to be there a month after us, so there was no point" (Inspector, focus group 4).

In principle, at least some integration of this kind can occur when an inspection and assessment are synchronized, but still conducted by completely separate teams. For example, inspectors and assessors could meet regularly throughout a synchronized engagement, exchange information on what they have observed to date, reach agreements on what each team should focus on going forward, and finally develop a mutual judgment on the performance of the activity examined during their visit. As explained previously, participants we spoke with during our fieldwork suggested that such coordination is unusual at best. They indicated that it is far more common for assessors and inspectors to share no information on substantive findings at all. At most, they coordinate their schedules so that their visits do not interfere with one another.

Participants told us that to changing this way of operating will require a new approach that begins with coordinated planning. Inspectors and assessors must agree on what items they will highlight and what perspective they will bring to each item. To keep things manageable, they must agree to split the workload and assign tasks to specific people they can rely on to execute the plan. This is where management gets more difficult. For example, what items on the checklist will each individual emphasize? Can each individual appropriately balance the priorities of the inspectors and the functional assessors? If not, are separate individuals required to support the priorities of each side? For example, one inspector with previous experience as an assessor noted how the jobs differed:

> It's one thing to inspect different slices of a program, and then to inspect it as a whole at the wing level. When I am there [for medical], I am looking at the wing perspective. It may be the same programs, but from a different angle. I look at the overarching process. They [other inspections] look at small details and I look at the big picture. You shouldn't combine [inspections with] different perspectives (Inspector, focus group 2).

Such initial planning might occur without formally assigning a lead organization. But once on site, having a single leader in charge becomes increasingly important. The leader can prioritize issues for regular interim reports to the unit under inspection. Also, the leader can adjudicate conflicts between inspectors and assessors and update priorities as additional information becomes available. In particular, the leader can facilitate effective data sharing. And, at the end on an inspection, the leader can ensure that inspectors and assessors generate a consistent message in response to each item on the relevant checklist.

Our participants voiced particular concerns about this last point. When a wing commander receives conflicting messages from inspection and assessment teams on site at the same time, it can leave the commander wondering which to favor or whether either is valid. Even after the considerable cost of preparation, inspection, and assessment, the potential exists for the effort to yield no feedback the commander is willing to rely on. As one inspector summarized, "It's embarrassing for IG to give an 'outstand[ing]' if LCAP gives an 'unsat[isfactory].' A commander will toss that report in the waste basket" (Inspector, focus group 1). Different grades can occur for the same performance when inspectors and assessors interpret checklists in different ways.

In general, interview and focus group participants strongly endorsed the value of integration, but they worried that it is difficult to successfully execute. To guide its implementation, the Air Force needs mechanisms in place to identify the relevant perspectives to each checklist item, establish how it should collect information, and how it should translate that information into a single grade for each activity that a unit commander will accept as valid.

Use Sampling to Reduce the Time Spent to Assess Any Activity During an Inspection Event

Any time an inspector, evaluator, or assessor draws an inference about the performance of an organization without observing every relevant aspect of that organization, some form of "sampling" occurs. For simplicity, we use "sampling" as a catchall phrase for any instance in which an inspector looks at less than the whole. Viewed in this way, sampling is obviously pervasive in Air Force inspection events. Inspectors never check the compliance of every part of an organization with every enforceable requirement in an AFI at all times. As TIG [The Inspector General of the Air Force] noted in his opening remarks at the October 2011 IG conference, there are roughly 200,000 compliance items for an Air Force wing. External inspections do not and cannot ensure compliance with all of them—or even most.[5] Such comprehensive oversight would not be feasible, much less cost-effective.

The simple act of inspecting at discrete points in time rather than continuously—of setting inspection intervals higher than zero—is a form of sampling. Similarly, the arguments we made in Chapter Two about the connection between the inspection interval and the governance structure of a local component of a broader organization illustrate a form of sampling.

This is true whether the sampling is precisely defined, completely ad hoc, or reliant on an inspector's intuitive ability to sense and track trouble by drawing increasingly narrow samples until a specific case of noncompliance becomes apparent. In fact, we can think of these various forms of sampling as alternatives from which an inspector can choose, using the framework we presented with regard to the frequency of inspection to identify the method best suited to a particular set of circumstances.

This section briefly revisits the framework presented in Chapter Two to show how it applies to sampling as well as the frequency of inspection. This framework provides a structured means through which to interpret how Air Force personnel in the field view the current use of sampling in the Air Force.

Sampling and Governance

The practice of sampling is among the many moving parts of the governance structure of an organizational component (for example, a wing) of a broader organization. This broader organization must coordinate the use of sampling with other elements of its components' governance structures. Like inspection frequency, sampling is a means through which the broader organization can convey information to an organizational component subject to inspection about how well its behavior aligns to the goals of the broader organization.

For reasons completely analogous to those explained in Chapter Two, we can make the following predictive statements about how the broader organization would behave if it wanted to integrate its sampling practices into a broader governance structure:

---

[5]   Rogers, 2011, Chart 7.

- Samples will be smaller—less broad and deep—when an activity to be inspected demonstrates that it has the local capability and motivation to promote the broader organization's goals in the absence of external inspection.
- Samples will be more focused on process than on outputs and resource consumption when an activity subject to external inspection has demonstrated that it is more capable and motivated.
- The size and focus of samples will change over time relative to the demonstrated capability and motivation of an activity under inspection.
- The size and focus of samples can appropriately differ for different kinds of external inspections of the same activity. One sample may be explicitly designed to prepare for a second as part of the general inspection plan for an activity.
- The more a serious negative outcome is likely to occur soon without prompt intervention, the larger—broader and deeper—samples will be.
- Samples will be larger—broader and deeper—for inspections of activities judged to be more critical to the performance of the broader organization.
- The broader organization's general desire for administrative simplicity will limit the degree of variation in formal sample design implied by the statements above.

We can expect two additional patterns that draw on similar logic, but do not have direct analogs to the choice of inspection frequency. First, samples will focus on those elements of an inspected activity that are of greatest importance to the performance of the broader organization. In an Air Force setting, for example, sampling will focus on the least capable personnel or parts of an activity under inspection if improving their performance is the most cost-effective way of improving the performance of the Air Force as a whole. If, on the other hand, the average performance of the activity is more important to the performance of the Air Force, sampling will focus on more representative personnel and parts of the activity. If inspections seek primarily to identify trends relevant to the general state of the Air Force, sampling can focus on selected wings rather than on selected parts of those wings. If, on the other hand, inspections seek to motivate all wings to sustain their performance, the sampling design must ensure that inspectors examine all wings, even if sampling is selective within wings. That is, sampling design depends fundamentally on the role that inspections play in the governance structures the Air Force maintains for its wings.

Second, more systematically designed samples can yield significantly more useful information about an activity under inspection than ad hoc samples can, but they may cost more to generate. External inspectors must be trained to apply formal sampling methods, and each individual design can be demanding for an inspector to prepare.[6] Cost-effective organizations balance these factors as they design their general policies on sampling. As external inspectors receive more information from new virtual inspection systems, the cost-effectiveness of formal sample design is likely to rise, allowing the use of substantially smaller samples. The desire for

---

[6] For example, formal sampling design might respond to guidance designed to avoid predictable forms of bias, to achieve some desired level of statistical power or confidence in applications of statistical inference, or to use the information collection resources available to drive down errors associated with particular estimates as much as possible. Such guidance would have to translate sophisticated statistical methods into practical terms that Air Force inspectors could apply with an appropriate level of confidence.

administrative simplicity could support the development of standard templates for application in a small number of circumstances.

Like inspection frequency, sampling design is part of the broader governance structure in a complex organization. For the Air Force, the size and focus of any sample depends on the organization's priorities, the local capabilities and motivation of activities subject to external inspection, the care taken in sampling design, and so on. Any change in one of these factors will alter the most desirable design of a sample. The desirability of any sampling size at any point in time depends on the state of the rest of the governance structure. In an inspection system that expects to change its basic approach to sampling over time, it is important to embed the plans for sampling into plans for broader change of the governance structure as a whole. It will be useful for all parts of the governance structure to move forward in a coordinated way. If not, significant changes in sampling design—especially reductions in size and changes in focus—could easily have unanticipated, negative effects.

How Air Force Personnel View Sampling in the Field[7]

Even when the people we spoke with did not use the word "sampling" to describe their actions, it was clear that they recognized the importance of focusing their attention where it is most important. This is second nature to the inspectors we met, who use the time and resources available to determine how deeply to sample. For example, one inspector said,

> We try to use valid sampling techniques. SAF/IG has guidance on what a valid sample is. Whether or not you sample 100 percent is left to the [inspection team]. If you have ten people [on an inspection team], you try to get it all crossed off if you can. If you can only get 25 percent, you sample that 25 percent (Inspector, focus group 4).

In practice, the inspectors' time and resources on site are determined well in advance. Because inspectors conduct so many inspections each year, the planning they can devote to any particular one is limited. Similarly, augmentees, by definition, assist inspections between other duties and can typically commit their allotted time only to the inspection in question. Once the time and resources available to execute an inspection task have been delineated, inspectors operate within those set bounds. The principal factor driving the size and shape of the samples inspectors draw is not the achievement of a certain level of confidence in the conclusions drawn, but rather the resources available to draw and assess a sample.

Inspectors learn how to draw these samples primarily on the job—they learn by doing. Formal Air Force guidance on sampling is very broad—too broad to inform the details of any specific sampling plan. While more detailed but less formal guidance is available, no one we spoke to mentioned this guidance.

Inspectors enter each inspection with the checklists relevant to that inspection. Often, the principal analysis they do to prepare for an inspection is to review the activities they will examine and determine, from their perspective, which checklists apply. They enter the inspection with the goal of collecting enough information to fill out those checklists. In fact, the formal checklists that Air Force inspectors use highlight only a fraction of the 200,000 or so items that AFIs state Air Force wings must comply with. Inspectors have the authority to call on any

---

[7]   All normative or prescriptive statements in this section reflect statements we heard from Air Force personnel, not RAND's independent judgment.

of these 200,000 items during the course of their inspection; but direct focus on the checklist significantly narrows the range an inspector will have to write about.

In effect, Air Force checklists already embody a significant set of sampling decisions. We heard concerns from our interview and focus group participants that the items included in these checklists are not always the most appropriate, however. Functionals at HAF compose Air Force–wide checklists. Functionals at MAJCOM headquarters add items specific to each command. Individual wings can add their own items to reflect wing-specific instructions. An inspector uses a checklist that reflects items identified at all these levels. As one inspector put it, "I just assess compliance, I do not make the checklist. . . . I'm not the legislator. I'm law enforcement" (Inspector, focus group 2).

Several of our participants believed that the experts who create many of these checklists are out of touch with realities on the ground. One inspector, for example, said, "because they have no experience [conducting inspections], the checklists/guides are too long for the time [allotted]" (Inspector, focus group 1). An inspectee agreed:

> You can't put 25 gallons in a five-gallon container. You need to keep it at five gallons. Instead of 4,000 checklist items, it should be 400. The functional commanders today are disconnected from the installation. Functionals today don't know how to take the 4,000 items down to 400 (Inspectee, leadership interview 4).

Another participant told us that, "For cops, there is a new HAF checklist, which went from 500 questions in the old compliance inspection list to 1,100 questions with the new HAF list. You need to either increase manning or note that not all questions are relevant to a particular unit" (Inspector, focus group 2). If neither occurs, inspectors must make sampling decisions.

The checklists do not reflect the priorities of individual wings: "There are some things in the guide that are not relevant. The things we are told to inspect aren't necessarily critical to the function" (Inspector, focus group 1). Functionals can also have difficulty understanding "inspectability"—how easy it is to judge compliance with the items listed in a real-world setting.

Perhaps most challenging of all, the checklists include more items than any inspector can reasonably expect to investigate and provide limited guidance as to which items the inspector should give the most attention. Lists of high-priority items and special-interest items give the inspector some guidance. This is where the inspector focuses the most attention. If time and resources are available to look beyond these, the inspector does so. In sum, higher-level organizations guide the inspectors' sampling decisions by highlighting a few items for special emphasis; the rest is up to the inspectors themselves.

Several people complained that augmentees tend to be more erratic. When augmentees rely too heavily on the practices at their home units to judge the units they inspect, the people being inspected tend to blame the absence of standard inspection terms. Perhaps a better way to understand this complaint is to appreciate that, because inspectors learn how to focus their attention through their hands-on experience, they develop their own standards over time. While augmentees might achieve something similar over time, they lack the experience of full-time inspectors.

Once on site, inspectors told us that they tend to focus their attention in a fairly predictable way. They begin at a relatively high level, surveying the territory they must cover for trouble. If initial probes in any area yield no surprises, they move on. This high-level survey can continue until they have initially tested all areas. If they encounter anything unexpected, they stop to examine it further. These unexpected observances include any failure to comply with a specific item of interest. One inspector told us:

> [I normally start with] the checklist. And it is more like a surface scan. The subset is based off of a surface scan. I'll look at the program book and safety records. If it gets to a point where I notice the documentation is not correct, I'll look further to see if it is this person or the system (Inspector, focus group 7).

Another said that if the personnel being inspected "are knowledgeable, we get through it fast. If they are not, that is when we spend more time" (Inspector, focus group 7). Inspectors continue to dig until they are satisfied that they understand the extent of the failure. One person called it "onion peeling," saying,

> For example, if I am looking for an inventory of 100 radios, [I'll pick a sample and ask for those particular radios.] If they can't find five [of the sample], I go to ten, if they can't find that, I go to 20, and so on (Inspector, focus group 7).

Inspectors are explicitly asked not to seek root causes of such failure. They leave this task to the wings in which the problems are found, but they continue to look until they find no further variances from expected standards. This probing is something like a dentist drilling to remove all of a cavity. The drilling does not necessarily reveal why the cavity developed, but it clearly defines the limits of its negative effects.

This two-step approach—a broad survey followed by exploratory dives where necessary—is a form of dynamic sampling. Initial data reveal where future sampling should occur. The first step of this approach necessarily skims the surface, choosing a small fraction of potential targets to test what might lie beneath the surface. When asked which items they tend to examine in this initial scan, inspectors gave us a broad range of answers.

In some cases, they look at the unit's previous assessments and inspections and give special attention to places where it did not do well in the past. One inspector told us that he puts "focus on areas where they were deficient before. Focus on where the improvement should have been" (Inspector, focus group 3). Another agreed: because "repeat write-ups are a death sentence for a unit . . . they are the first place to start" (Inspector, focus group 3).

This approach, however, is apparently not common, at least not among our focus group participants. The majority of inspectors we spoke with prefer to come to a unit with an open mind and therefore avoid such past performance information. More often, an inspector thinks about where s/he has seen weaknesses in other units with analogous activities. If s/he has seen a pattern of weakness in a particular area across units, s/he gives that area special attention in the next inspection. One inspector said, "We look for trends. . . . It is more important . . . to inspect based on priorities and trending problems than it is to go through the checklist line by line to waste time looking at wires behind a computer that have never been a problem in the past" (Inspector, focus group 8). Another said, "If other programs [at other bases] are having problems, but you say your program is doing fine, then I want to know why yours is doing so well, or I need to dig deeper and find out if yours is really doing fine" (Inspector, focus group 1).

Others highlight certain areas to which they regularly give special attention. In addition to the MAJCOM high-priority areas and special interest items, various inspectors highlight areas with direct effects on personnel safety and mission effectiveness, and areas that affect assets or activities with a high dollar value.

Within these areas, inspectors offered distinct strategies. Some seek the places where weaknesses are likely to be largest. For example, they ask to see the records of personnel having the greatest trouble progressing through training milestones and ask what a wing is doing to manage each case. One inspector said,

> Seventy percent of the students that go through are fine. You want to focus on the students with the problems, focus on those in the elimination process and what they are going through and what the commander's review process is for them. You don't want to focus on those that can take the book and self-learn. You sample 100 percent of those, because those [failing students] are where the commander gets into trouble (Inspector, focus group 4).

Recruiting is challenging and costly. Commanders seek to retain as many people who have been successfully recruited as possible.

Other inspectors seek a representative cross section of items to sample. They ask for a sample that includes high-, low-, and average-scoring performers, then examine this sample in more detail. Or, in an organization that manages many different kinds of equipment or records, inspectors try to see examples of each type. One inspector said, for example,

> We should be looking at all flight record folders. But there can be 1,700 records. So even a 10-percent solution is more than we can do. So we take . . . a sample of one of every type of aircrew assigned to the base (Inspector, focus group 9).

Another said, "In Fire, we . . . pick one person per shift, over each shift which is eight hours. I'll look at a level 3 person, a level 5, and a level 7 per shift" (Inspector, focus group 4). This approach is presumably more motivated by a desire to understand the overall status of a unit than a desire to identify the worst performance.

In effect, these two approaches to sampling define qualitatively different metrics of performance.[8] Inspectors looking for the worst performers will apply a maximin standard—the higher a unit's minimum performance in any area, the higher it scores. Inspectors looking for representative performers will apply a standard based on a central tendency—the higher a unit's average or median level of performance in any area, the higher it scores. These two standards lead to very different behaviors in a unit subject to inspection. Our interview and focus groups did not give us enough information to verify that inspectors intended to induce these different behaviors by applying these sampling choices. It was not clear from our discussions whether the Air Force maintains any guidelines—formal or informal—for imposing different types of standards in different circumstances.

A third approach focuses on data from prior to when a unit began preparing for inspection. For example, if a particular inspection occurs every four years or so and the inspector believes that a unit spends nine months preparing for an announced inspection, the inspector ignores records from those nine months and focuses instead on records that are nine to, say, 24 months old. Looking at data more than 24 months old can be difficult because stan-

---

[8]   For a discussion of the broader implications of such a choice, see Stecher et al., 2010.

dards change and a unit cannot be held accountable for standards that were not in place at the time the data were recorded. One inspector called this period nine months out the "IG coma period"—the period during which the expectation of an inspection is not actively on the minds of personnel in the unit. This approach seeks to emphasize the actions taken by an organization when it is not focused on inspection preparation. Looking far enough back to do this takes the emphasis off the current leaders of the unit, however, and is not a reliable way to judge their performance and hold them accountable for it.

Use a Higher Percentage of No-Notice Inspection Events

In the Air Force, "no-notice" inspections are those in which a wing receives 72 hours of notice or less prior to being visited (AFI 90-201, 2009). This short notice is intended to be sufficient for the wing to manage logistics for the visiting inspectors, and for inspectors to coordinate their visit with wing activities without seriously disrupting the execution of their primary missions. To step on base unannounced without seriously disrupting the base, such inspections must be much more limited in their scope than standard compliance and readiness inspections. Despite this, the few days of preparation time do not give a wing any chance to "get well" before the inspectors arrive unless the wing is already performing at close to its nominal goals. As a result, advocates of no-notice inspections argue that they encourage wings to operate close to their nominal goals at all times. Also, the smaller size and scope of the inspections limits the cost of achieving this high and persistent level of wing performance.

This section briefly revisits the framework presented in Chapter Two, showing how it applies to no-notice inspections as well as the frequency of inspection. This framework provides a structure for interpreting how Air Force personnel in the field view the current use of no-notice inspections.

No-Notice Inspections and Governance

Like inspection frequency and sample design, a no-notice inspection is one more element of the governance structure the Air Force uses to align the behavior of individual wings to Air Force–wide goals. Consequently, the design and application of no-notice inspections are subject to exactly to same considerations we have already discussed with regard to inspection frequency and sample design. Using the same logic we have applied twice above, we offer the following statements about no-notice inspections:

- Inspections can be smaller, less frequent, and more cost-effective when the activity to be inspected demonstrates the local capability and motivation to promote the broader organization's goals in the absence of external inspection.
- Inspections can be more focused on process than outputs and resource consumption when the activity subject to external inspection has demonstrated that it is more capable and appropriately motivated.
- The size and frequency of inspections can change over time relative to the demonstrated capability and motivation of the activity under inspection.
- It could be cost-effective to use a no-notice inspection to determine the desirability of a more comprehensive and time-consuming inspection. The no-notice inspection might suggest where a follow-on inspection should place greater emphasis.

- No-notice inspections can become larger and more frequent when problems are likely to get worse quickly.
- Inspections can be larger and more frequent for those activities judged to be more critical to the performance of the broader organization.
- Subject to all the considerations above, no-notice inspections can be smaller and more frequent if they are better designed to collect relevant information at a low cost to inspectors and the units inspected.
- A broader organization's general desire for administrative simplicity will limit the degree of variation in the size and frequency of no-notice inspections implied by the statements above.

The general themes applied perviously to inspection frequency and sampling design apply equally well here. No-notice inspections are simply additional moving parts in the Air Force's governance structure and are therefore more likely to produce their desired effects if coordinated with other parts of the governance structure for any wing.

How Air Force Personnel View No-Notice Inspections in the Field[9]
We found general support for no-notice inspections. Indeed, interview and focus group participants often broached the subject of no-notice inspections and noted their benefits without prompting. One participant emphasized how no-notice inspections would induce a different kind of culture: "Having ownership . . . [h]aving integrity, using correct judgment, being held accountable with a surprise inspection, I think that would be much better" (Inspectee, NCO focus group 1). Another NCO in the same session asserted that the current inspection system is "training airmen to fake it." Similarly, a participant in another focus group remarked,

> No-notice would make the inspections more effective. With the UCI as it is now, you put all the junk on your desk into a drawer while the IG comes around. Then when they leave, you pull it all out again and go back to normal (Inspectee, officer focus group 1).

But we also heard that, to be practicable, such inspections would have to differ in important ways from the general compliance inspections the Air Force uses today. Current experience with LCAPs helped inform a number of people we spoke to. LCAPs give only 45 days of notice before an assessment, far less than a standard IG compliance inspection, but many respondents spoke of the value of giving as little as 48 hours of advance notice.

Our interview and focus group participants liked the idea that, because they could occur at any time, no-notice inspections would probably increase the general level of compliance in wings over long periods of time. This view flows from the broadly held belief, discussed above, that compliance inevitably falls off between scheduled inspections and falls off more the longer the interval between inspections. If inspectors are not watching, wings will shift their priorities to other things. As one focus group participant put it, "The only way you'll know how the wing is doing is if they don't have notice that [the IG is] coming. If you stand off and they don't know you're there, you get a better picture of how they're doing" (Inspectee, NCO focus group 4). The use of no-notice inspection dramatically reduces the potential time interval

---

[9] All normative or prescriptive statements in this section reflect statements we heard from Air Force personnel, not RAND's independent judgment.

between inspections and thereby encourages wings to stay in compliance at all times. As one NCO told us, "That will drive compliance all the time. It will make sure there are teeth in my self-inspection program" (Inspectee, NCO focus group 2).

Participants also liked the idea that, if a wing could not know when a no-notice inspection might come, it could not and therefore would not have to put as much effort into preparation as it does now. One airman said,

> That's how they do it for radioactive material. They just show up on base. It is awesome; I never have to prep. If I am doing a bad job, they will find it (Inspectee, officer focus group 3).

Note the potential for a contradiction here. If a wing always remains prepared for a no-notice inspection that could come at any time, it is hard to understand how this requires less effort than preparing only periodically. There is a sense that no-notice inspections might offer the Air Force something for free—better compliance without the effort required to ensure compliance.

When pressed to explain how this might be feasible, participants typically suggested that current inspection preparation is excessive. To ensure the best grades available, wings tend to shine themselves up in ways that are not really relevant to meaningful compliance with AFIs. One inspector told us, "we spend lots of time and energy trying to get through the 'polishing' that goes on" (Inspector, focus group 3). Another inspector in the same session believed that "units waste time getting 'eye candy' for inspections." As one airman put it, "Many units create superfluous products, programs, eyewash for the inspectors that will then collect dust for two years" (Inspectee, leadership interview 10). Units may offer elaborate hospitality to visiting inspectors to win their favor, prepare elaborate briefing books to impress them, or paint facilities that do not require painting. Our participants suggested that, because such elaborate preparation simply could not be sustained at all times, the use of no-notice inspections would discourage it, allowing wings to find a better, more sustainable balance between their compliance and mission priorities. As one person put it, "It forces you to maintain programs, none of this prep" (Inspectee, officer focus group 3). Another participant in the same session said, "It limits polishing aspects—a lot of work for nothing."

One person noted that getting to a sustainable balance would be worthwhile, even if the journey there proved rough:

> If they do no-notice, [the IG] will see the reason that people put stuff on the back burner. They will be able to do trend analysis and find out what's not important, and then they can take it out. . . . [Currently, units] are putting in all these man-hours and [the IG] is saying, "Well, people are able to do this; you are just not working hard enough [if you fail]." But if ten bases in a row fail, that would help AFSO21. . . . Maybe they will give me more man-hours (Inspectee, officer focus group 3).

Other participants suggested that no-notice inspections would have to differ from current inspections in specific ways to encourage such an improved balance. Perhaps most important, the Air Force would have to move from a five-tier grading system to something closer to a pass-fail system, in which a unit is either in compliance or not. One person suggested,

You can't have satisfactory compliance. Compliance is compliance. 8 PSI [pounds per square inch] for a tire is compliance. Having a shiny wrench makes the unit "compliant," not "outstanding" because it is shiny (Inspectee, leadership interview 4).

According to this argument, much of the excessive elaboration in current preparation is the result of the efforts of wings to achieve the highest grade possible. A system closer to pass-fail would not reward such effort and focus a wing on achieving only the level of performance required to be in satisfactory compliance with AFIs.

One participant made a closely related recommendation. He suggested that current inspections induce wings to look as sharp as possible when an inspection is scheduled. Because no organization can maintain such a sharp edge at all times, inspectors in a no-notice inspection would have to accept this and assign grades in a way that did not penalize them for it. As one person put it,

These units are paced to reach peak performance for the inspection. I don't know any sports team that keeps at peak for the duration without getting worn out. The units are always at really high-tempo prep (Inspectee, leadership interview 9).

No-notice teams would also have to be smaller. As discussed above, the logistics of a major inspection place a heavy burden on all but the largest wings. The requirement to house and feed inspectors, provide vehicles and facilities, and divert personnel from ongoing mission activities requires significant planning. Without the opportunity to prepare, inspections of the current magnitude would simply overwhelm most wings. If inspection teams became smaller, this would create the additional benefit of reducing the local burden of an inspection whenever it actually occurs.

Finally, inspection teams would have to accept that, without the time to prepare properly, wings would necessarily offer less elaborate accommodations. As one person told us,

Don't expect binders. We will place you in lodging wherever we can. We won't greet you at the gate. Don't expect to be fed at the start. Don't take that out on us. We could give binders to inspectors who came for our no-notice HSI, but that's only because we had already prepped the material for the UCI (Inspectee, officer focus group 1).

In effect, inspection teams would have to lower their expectations and learn to live with less comfortable circumstances than they have come to expect without penalizing individual wings for that change.

Summary

Synchronization and integration of inspection events, sampling design, and no-notice inspections can potentially be coordinated as elements of the governance structure the Air Force uses to align its wings under broader Air Force goals. They are all elements of the governance structure that generate information about how well a wing is aligning its behavior and performance to broader Air Force goals. Also, ongoing efforts to reduce the cost of the Air Force inspection system focus special attention on these elements. For those reasons, we have discussed them together in this chapter, but they are only a few of the many moving parts in the governance structure of a wing.

Most personnel in the field favor synchronization, which is already well underway in the Air Force, but they want synchronized inspections and assessments to be better integrated than they typically are now. Effective integration is challenging. To achieve it, someone will have to go through every checklist item to decide (1) which items are really worth tracking and (2) whether the perspectives of both an inspector and a functional assessor are necessary for each item. This task will require greater coordination between the IG and the functional communities than has been typical in the past.

Sampling is a pervasive practice in the Air Force inspection system, but it does not appear to be informed systematically by formal training or the clear prioritization of issues relevant to oversight. The results of our fieldwork suggest that inspectors learn how to sample primarily on the job. Moreover, the credibility of the current inspection system rests heavily on the sampling skills and discretionary decisions of experienced inspectors. We found broad support for the increased use of no-notice inspections, but interview and focus group participants expect such inspections to work only if external inspection teams (1) are much smaller than those traditionally used, (2) temper their expectations about performance, and (3) potentially move from the current five-grade approach to a pass-fail approach to scoring.

The local capabilities and motivation of each wing are also critical parts of its governance structure. The next three chapters address a few elements of local capability and motivation that are relevant to ongoing change in the Air Force. Chapter Four discusses increasing the emphasis placed on self-reporting relative to the external inspection of wings. Chapter Five discusses how to measure the quality of leadership and discipline at a wing. Chapter Six discusses full diffusion across the Air Force of MICT, an information system the Air Force is using to increase local capabilities.

The better coordinated these changes to the Air Force inspection system are, the more cost-effective that system is likely to become. As changes proceed, the Air Force should ensure that each change creates conditions that support every other change. Implementing many changes at once will prove challenging. While the Air Force cannot know exactly how to coordinate these changes in advance, it can learn as implementation proceeds by monitoring changes and adjusting them over time to improve their coordination. The next three chapters will say more about what this entails.

CHAPTER FOUR

Shift in Relative Emphasis of External Inspection and Wing Self-Reporting

To ensure readiness and compliance in its wings, the Air Force currently relies primarily on having inspectors, assessors, and evaluators from outside the wings monitor their performance and advise them on how to improve it. As a complement to this independent, external perspective, wings have a variety of internal self-inspection practices, but skepticism currently exists about whether these generate the information required to sustain reliable performance. At the time of this report, SAF/IG is laying the groundwork for a major change that would further emphasize self-inspections and reduce external oversight.

This chapter reports our findings on Air Force perspectives regarding this inspection system paradigm shift. It also presents information on another way to shift from external oversight toward the bottom-up reporting used by the FAA. The Air Force is experimenting with an approach similar to the FAA's and could dramatically expand its application. This chapter closes with a summary of lessons learned for the Air Force.

Greater Reliance on a Wing's Own Self-Inspection Practices

According to AFI 90-201, self-inspection "provides commanders with a tool for internal assessment of unit health and complements external assessments" (2009, p. 16). Each MAJCOM sets its own self-inspection guidelines, but essentially wings use HAF checklists—and other checklists as appropriate—to monitor for instances of non-compliance and related deficiencies. Once shortcomings are identified, root cause analysis and corrective actions are completed at the lowest possible level of responsibility. The Air Force Smart Operations for the 21st Century (AFSO21) Playbook is identified as a key source of guidance for such problem solving.

The process of self-inspection is discussed in AFI 90-201 as one way of reducing the inspection footprint. Specifically, MAJCOMs may give units "credit" for their self-inspection activities by reducing the nature and/or extent of external inspections. One change to the inspection system that the TIG and the ISITT have opted to pursue involves taking this practice further: new guidelines have been developed for a rigorous, robust CCIP that, once fully implemented, will result in a greater reliance on the wing's own inspection practices and a reduced emphasis on (though not an elimination of) external inspections.

In this section, we discuss findings from our fieldwork that relate to this shift in inspection responsibility. Specifically, we summarize focus group and interview participants' views on the possible benefits of this change, potential concerns, and ways to implement the shift successfully. We also draw on literature related to psychological safety to offer additional implementation-related guidance. Finally, we provide examples from current inspection preparation practices that demonstrate that the underpinnings of the new CCIP are, at least to a limited extent, already in place.

Perceived Benefits of Greater Reliance on a Unit's Own Self-Inspection Practices

During our focus groups and interviews, we presented participants with a potential scenario, one in which there was a greater reliance on wings' self-inspection practices and less emphasis on external inspection, and asked them what the pros and cons would be of this type of change. Neither the inspector focus groups nor the inspectee focus groups and interviews gave many specific comments about benefits of this type of change. Instead, positive remarks were of a vague or general nature, such as "That is an excellent idea" (Inspectee, NCO focus group 2), or "Overall, I like the idea, but I would caveat it" (Inspectee, officer focus group 1). Similarly, an inspector told us, "If units are doing self-inspections annually, they might as well do that and let it replace the compliance inspection" (Inspector, focus group 3). This may have been because participants were thinking about the current system of wing self-inspections and compliance inspections, rather than the changes being implemented at the time of this report—namely, the move toward the UEI and CCIP.

Concerns about This Potential Change

While there was a dearth of evidence to support the benefits of a greater reliance on units' self-inspection practices, our interviews and focus groups generated numerous warnings about the potential change and its perceived drawbacks. Such comments typically related to the varied quality of wing self-inspection practices or the lack of standardization, two issues regarded as hindrances to effectively increasing emphasis on wing self-inspection across the Air Force.

Perhaps because of their exposure to an assortment of wing self-inspection programs, the inspectors we spoke with were especially inclined to comment on how much their quality varied. As one inspector put it,

> There is a complete 180 on self-inspections. I have seen "yes, yes, yes, yes," pencil-whipping-type self-inspections. I have also seen ones where the question has been researched, the history is shown and tracked, and leadership has signed off on it. They vary on a 0–10 scale (Inspector, focus group 5).

Remarks from both inspectors and inspectees offer insights into why self-inspection quality tended to differ from wing to wing. First, there was a sense that commanders may prioritize self-inspection differently given their available resources, as illustrated by the following comments:

> That comes down to the commander's priorities. The Air Force says operations are important, need bombs on target, so not worried about other documentation and whether it's perfect. One of the fallbacks of self-inspection is that the local commander will prioritize. Some will be willing to take the hit [sacrificing compliance for mission completion] (Inspectee, officer focus group 2).

> At the end of the day, we are focused on cargo packs and airplanes in the air. In the day-to-day we need to do a core set of items. Voting guides and training guides? No unit gives

that stuff so much emphasis unless an inspection is coming up. But, I do weapons every day (Inspectee, officer focus group 1).

Also, inspectors, personnel from inspected wings, and leadership from inspected wings frequently suggested that, while airmen may be well qualified to perform their primary duty assignment, they may not possess the qualifications needed to serve in an inspector role:

[A]cross the Air Force, the people responsible for preparing the wing are not in the same office. In this wing, the Plans and Programs office leads the prep for inspections. In other Air Force units, it's the wing IG office. So it's not standardized who in the wing will lead that effort. In that office [Plans and Programs], it's usually aircrew guys who spend a short time, six months, in the office to punch a ticket and then leave. A lot of people there are not trained. You depend on them, but this is not their core job (Inspectee, NCO focus group 2).

People don't necessarily know what a good self-inspection looks like (Inspector, focus group 6).

[F]inding deficiencies and correcting them is not about just a point of time but making a better process, and if you are not trained or in that mindset, it is not going to be a practical tool (Inspectee, officer focus group 3).

I don't think they [self-inspections] should be done at the squadron level. That's too localized. You need to have them at a level where you know the expertise is competent. We didn't have the expertise at the squadron level since the Air Force has pulled the programmatic expertise out of the squadron. There's no dedicated training guy, safety/security guy, etc., at the squadron. Since you don't have dedicated personnel to do it, people should get training outside of the squadron, and do self-inspections at the wing level. No lower based on the current lack of manning power (Inspectee, leadership interview 6).

A final explanation that emerged from the inspector and inspectee focus groups was that, when it comes to self-inspection, not all units will be honest with themselves. As one inspector put it, "You have to be brutally honest in an SIP [Self-Inspection Program]. Some are, some aren't" (Inspector, focus group 1). This could stem from naiveté, a lack of knowledge, or, as the remarks that follow suggest, a lack of willingness either to admit mistakes or present evidence of non-compliance to one's colleagues:

A unit's Standards Evaluation Unit is like a mini IG. You need strong people willing to write up their own unit, but there's still self-preservation [involved]. You [Airmen] won't be in the Standards Evaluation Unit forever. It is extremely difficult to police yourself and honestly say, "We messed up" (Inspector, focus group 2).

Self-inspections go up the chain of command. Nobody is going to say, "We are terrible. I need to be fired." It is kind of a self-decapitation. Some individuals will, some won't (Inspector, focus group 9).

These comments suggest that wings may provide their members with too little "psychological safety." Psychological safety is a shared belief held by members of a group or organization that the unit is safe for interpersonal risk taking (Edmondson, 1999). Conditions in which psychological safety is high may be conducive to the voluntary reporting of errors and a more proactive approach to self-inspection. According to Schein (1993, p. 89), elements of a psychologically safe environment include (1) training and practice opportunities, (2) encouragement and support to alleviate concerns associated with committing errors, (3) rewards and coaching for efforts "in the right direction," (4) norms that establish that committing errors may be permissible under certain circumstances, and (5) norms that reward innovation and creative thinking.

These elements likely are under the control of a unit's leadership. Indeed, more recently, Edmondson (2011, pp. 52–53) identified ways in which a leader can build a psychologically safe environment, such as framing work accurately, implementing blameless reporting, and holding people accountable for "blameworthy" acts. This suggests that it would be helpful to measure the ability of a wing's leadership to sustain the degree of psychological safety required to overcome the concerns our fieldwork participants voiced. Chapter Five will address this idea in more detail.

A separate line of discussion regarding the inadequacies of current self-inspection practices pertained to a lack of standardization across wings. This perceived variation in approach to self-inspection was seen, in part, as a consequence of AFIs that are generic or "written in gray." Bases and wings develop individual instructions that are consistent with these AFIs, yet take into consideration MAJCOM guidance as well as location-specific nuances.

Moreover, the large number of checklists that apply to wing operations makes it even more likely that what they include in their self-inspection programs will vary. As mentioned in Chapter Three, a contractor estimated that roughly 1,150 AFIs applied to the wing level, and that there were over 200,000 compliance items to inspect.[1] In a related vein, a leader from one of the recently inspected wings we studied told us,

> [T]he number-one problem we have is the fact that the self-inspection checklists are organized functionally instead of organizationally. If the contracting squadron commander could go and pull up the contracting squadron checklist, we would be there. Now, he has to go pull the contracting squadron functional checklist, then he has to read every other of the 36 topic areas [related to his work] to see if there is something else in there that applies to him. The group commander made me read every single thing in every single area [in prep for the UCI], and I discovered 30 checklists in the Mission Support Group (MSG) area we didn't even know about.
>
> **RAND: There are a large number of checklists that you have to find on your own?**
>
> I discovered a unit that went from having 30 to having 400 checklists. That's just from sitting here at night reading the checklists. There are items that touch every single unit on the base that they [those units] don't know (Inspectee leadership interview 12).

---

[1]   Rogers, 2011, Chart 7.

Although customizing self-inspection practices may naturally occur given the diversity in wing-level missions and the large number of applicable checklist items, it was seen as a potential challenge to the increased emphasis on self-inspection throughout the Air Force. Fieldwork participants recognized this as essentially a trade-off between specialization and standardization, one that currently limits the Air Force's ability to compare the results of self-inspection programs across wings.

Benefits of an External Look

While the preceding concerns were about the perceived shortcomings of current self-inspection practices, others pertained to losing the benefits of external inspection. Both inspectors and inspectees felt that retaining an external look was critical and cited several arguments in support of their opinion. The objectivity and fresh perspective that outsiders bring to a wing were considered important benefits to preserve within a new inspection system. As one inspector stated concisely, "The IG is objective. We don't care if they do well in the inspection—but we hope they do" (Inspector, focus group 7). In addition, as the comments below attest, a fresh set of eyes helps to identify deficiencies that an insider might miss:

> [I]t's useful to have someone from outside take a look. You get used to seeing the forest and you don't always see the trees. So it's useful to have someone else who's not familiar and get them to start asking questions (Inspectee, leadership interview 7).

> Even pro football players practice the same way and have their movements videotaped and others watch them to point out ways they can improve. [External inspections] can confirm if/when we are missing the little things (Inspectee, NCO focus group 3).

> I do internal QA [quality assurance], and I don't catch things. I do external inspections on other bases. Like being in your own house for a while and not noticing a painting until someone else points it out, being here for a long time, it is my backyard. I don't catch [problems] until I go somewhere else (Inspectee, NCO focus group 1).

An outside perspective was viewed as helpful not only to identify specific instances of noncompliance, but also, as a commander told us, to reveal program-level shortcomings:

> The self-inspection does not have any vector corrections. If the self-inspection moves, over time, off the center and in a certain direction, there is no outside look [without external inspections] to reset that directional vector to the center. Our self-inspection is a good example. Ours was way off, but we didn't know before having to prep for the UCI. Without outside guidance, we will just be deriving our own standards (Inspectee, leadership interview 2).

An external look was also valued because IG inspectors, or those from other outside authorities, are able and willing to be the bad guy. One commander observed, "[P]olicing yourself produces some problems" (Inspectee, leadership interview 7). Whereas an inspector told us, "As a job, we go in [to a unit] to call things ugly if we have to" (Inspector, focus group 8).

Earlier in this chapter, we noted that participants believed wing personnel's lack of inspection expertise contributed to disparities in self-inspection program quality. In a related vein, both inspectors and inspectees suggested that a key benefit of having an external look was the involvement of qualified inspectors possessing all the necessary expertise: inspection training, functional knowledge, and an up-to-date awareness of new regulations, instructions, and other guidance changes. Individuals who regularly conducted inspections were particularly valued as a source of "cross-tells" (sharing best practices and other lessons gleaned from visiting many wings) and were believed to be better inspectors as a result:

> There are also gains from our experience of repeated inspections. If other programs (at other bases) are having problems, but you say your program is doing fine, then I want to know why yours is doing so well, or I need to dig deeper and find out if yours is really doing fine (Inspector, focus group 1).

Another perceived benefit of an external look was voiced during our interviews with commanders, who felt that resource constraints may not be addressed until an external inspection documents their consequences:

> When resources are constrained, and we are trying to tell Higher Headquarters . . . [w]hen a major inspection and outside source comes in and they say the same thing [that we are resource constrained], that sometimes gets additional support. It validates that the complaint is right (Inspectee, leadership interview 9).

> It helps when we are trying to get new things, new training. It gives us a formal process to do that (Inspectee, leadership interview 4).

Suggestions for Implementation

During focus groups and interviews, we asked participants for suggestions on how to address the concerns outlined above and effectively increase the emphasis on self-inspection practices. Their responses ranged from positive and supportive to those more closely resembling negative reinforcement. On the positive end, Air Force personnel recommended that the Air Force add sufficient resources so that a wing need not resource self-inspection within its pre-existing budget and set of personnel (take it "out of hide"). Resources tended to be discussed in terms of manpower, as the following remarks demonstrate:

> You could create people whose job is self-inspection. It can't be an additional duty. If you create manpower billets and dedicate personnel to do that, then you might have a chance to make them successful. It probably takes a lot more manpower to do it in a decentralized way than a centralized IG (Inspectee, leadership interview 1).

> You'd have to do it like Ops does with Stan/Eval. Have a guy whose standing duty it is to inspect. Otherwise, you will get a guy who doesn't want to write someone up because it ruins his day-to-day interaction, but if that is his job, people know (Inspectee, officer focus group 3).

This suggestion was motivated in part by the belief that areas with dedicated, trained QA personnel had more robust self-inspection practices. This often occurred in response to outside pressures from federal law, the Occupational Safety and Health Administration (OSHA), and other agencies. As one focus group participant told us,

> Some squadrons are inundated day-to-day with compliance work versus others that don't have that external look. When there are external agencies, people tend to carve a piece of the program out for compliance. . . . We have to have [compliance] or else we get shut down. Without external agencies, [the squadrons] don't have to have internally built compliance (Inspectee, NCO focus group 1).

Another suggestion for ensuring compliance once self-inspection is given greater emphasis and IG teams visit less frequently is to invite other types of outsiders to give an objective, fresh look. These outsiders could come from another part of the wing, from another base, or from the MAJCOM. Retaining some sort of external look, even informally, was regarded by our participants as an important element of a self-inspection program, especially in light of decreased attention from IG inspectors.

Additional suggestions were focused on accountability. Focus group and interview participants felt that requiring documentation for self-inspection checklist items (i.e., a yes/no answer was not sufficient), auditing specific self-inspection results, and requiring that self-inspection results be forwarded to MAJCOMs were all ways of ensuring that self-inspection "had teeth"—that is, induces effective compliance. A final idea was to ensure there be consequences for a poorly executed self-inspection. This was offered as a way to hold a commander accountable. As one commander explained during his interview,

> I think that you can rely on self-inspections to a certain extent, but there has to be a consequence [if they are inaccurate]. If you're only relying on that [self-inspection results], then it becomes a paper tiger. What gets measured gets done (Inspectee, leadership interview 10).

Evidence from Wing Inspection Preparation Activities Suggest That a CCIP Is Feasible

As noted earlier in this chapter, personnel we spoke with were focused on the compliance inspection and self-inspection practices and programs in place at the time of our fieldwork (spring–summer 2011). However, the new inspection system paradigm includes a more robust self-inspection program, the CCIP. The CCIP "will coordinate wing-level inspections, assessments, evaluations, exercises, observations, and other measurements into a single, cohesive program focused on the commander's objectives and mission" (Hyde, 2011b, p. 1). The wing commander, vice wing commander, and wing IG will all play critical roles in ensuring the effectiveness of the CCIP, which will be validated by the MAJCOM IG inspection team.

The CCIP will make use of internal programs and activities including, but not limited to, the wing's QA program, wing Standardization Evaluation (Stan/Eval) program, exercises conducted by the wing's Exercise and Evaluation Team, and the wing's self-inspection program. External data sources, such as functional and staff assessments, will also inform the CCIP. While these efforts are more extensive than the efforts currently expended on self-inspection programs, evidence from our fieldwork indicates that some wings engage in these activities—and more—in preparation for the current compliance inspections. When asked how their wing got ready for their recent compliance inspection, many focus group and interview participants talked about using databases and other IT-based tools to develop, execute, and track compliance-related checklists. They also discussed setting up internal Tiger Teams consisting of experts from across the wing, and arranging visits for staff and personnel from other bases to check on the wing's inspection preparation progress and verify compliance. In addition, wings engaged in both function-specific and wing-wide compliance exercises. Finally, wings sought to learn compliance-related lessons, both good and bad, from other bases. Personnel reviewed recent inspection reports from other locations and, as resources permitted, traveled to other bases to serve as inspection observers. These varied activities helped to ensure that wings were as prepared as possible for the IG inspection team's visit, and, as one commander told us, also served as team-building exercises. Thus, requiring wings to formally use such programs and activities may not be a drastic departure from those practices currently in place within wings. Instead, such a mandate will ensure these programs and activities are consistently used by all wing commanders and may even offer additional benefits. Indeed, the CCIP is intended to improve effectiveness, readiness, discipline, and surety, and, given the observation about team building shared above, may also boost motivation and morale.

Federal Aviation Administration Voluntary Reporting Programs

The FAA has developed a sophisticated inspection system for the U.S. commercial aviation system that has had to address many of the same challenges the Air Force inspection system faces with respect to self-reporting. The FAA's use of voluntary reporting programs could interest the Air Force for two reasons. First, it offers a source of data that complements inputs from a more traditional inspection system focused on compliance. In particular, it offers data relevant to the future performance of an aviation system that a compliance-focused system cannot capture, since compliance standards cannot be written to address all events that have yet to occur. Voluntary self-reporting generates information that can potentially shape the new standards the Air Force will enforce in the future. Second, it directly addresses the challenge of encouraging knowledgeable personnel to report the negative information needed to diagnose problems in spite of their fear that such information could be used to punish them.

This section provides an overview of FAA voluntary reporting systems. It first describes the current FAA programs,[2] then briefly reports FAA perceptions of their costs and benefits, and outlines what steps the FAA took to create an environment with high psychological safety. While we cannot assume that the Air Force could or even should adopt an identical system, we offer information from the FAA to show what has apparently worked in another complex and challenging inspection setting. The section closes with a description of a relatively new Air Force program that emulates the FAA's programs and appears to be meeting initial success.

Current FAA Voluntary Reporting Programs

The American commercial aviation supply chain uses a system of tightly interlocked programs to collect extensive voluntary reports relevant to system safety from FAA air traffic controllers and FAA certificate holders—the air carriers, repair stations, parts manufacturers, and other private-sector organizations that the FAA regulates and certifies. The following voluntary reporting programs lie at the heart of this FAA-coordinated system: Aviation Safety Reporting System (ASRS), Aviation Safety Action Program (ASAP), Voluntary Disclosure Reporting

---

[2]   Chapter Seven summarizes the evolution of the FAA programs discussed here since the 1950s and the methods the FAA has used repeatedly to implement fundamental changes.

Program (VDRP), and Flight Operational Quality Assurance (FOQA). Voluntary programs within certificate holders support this system. Together, all of these voluntary reporting programs feed the Air Transportation Oversight System (ATOS) certification process that the FAA uses to certify its certificate holders.

The following attributes of this system of systems are potentially relevant to the Air Force inspection system and, in particular, the Air Force's use of data from wing self-inspection programs:

- The system gathers and integrates both self-reported and external observation data from multiple, independent sources, both objective and subjective.
- Voluntary reports account for a significant portion of the self-reported data collected. The system is explicitly designed to protect sources of voluntary reports from punishment to encourage true reporting on negative conditions.
- Voluntary reports are used primarily to *diagnose* problems in the aviation system as a whole and fix them before they precipitate accidents or other negative incidents, not to *motivate* individual parts of the system to perform well.[3]
- The system uses experts to assess new data continuously in the context of historical data to identify new patterns used to redefine formal compliance programs.

The FAA system of systems focuses on safety, but all its elements are equally relevant to broader questions of performance in the Air Force as a whole.

The Aviation Safety Reporting System is an industry-level system that receives, processes, and analyzes voluntary safety incident (near-miss) reports from pilots, air traffic controllers, dispatchers, flight attendants, maintenance technicians, and other stakeholders who directly affect commercial aviation operations and safety. ASRS receives about 50,000 reports per year, 60 percent of which are from pilots. The system has accumulated over 34 years of data. It currently has 1,200 data fields associated with the incident reports, which are kept and coded. Like other voluntary reporting programs, ASRS maintains the data collected from the incident reports in a narrative, qualitative format based on, by design, loose guidelines.

Subject matter experts, who are typically part-time contract government employees and commercial aviation industry retirees with over 25 years of experience in various areas, analyze these data to understand how the aviation system affects specific elements of human performance and human errors. The SMEs use these data to generate watch lists, alerts, emerging trends, and special studies that they distribute via alert bulletins, callback messages, and other methods to relevant stakeholders, including the other programs described below.

The FAA funds the ASRS, but it is actually administered by NASA's Human Systems Integration (HSI) group to maintain its independence from the FAA. NASA's role is designed

---

[3] Metrics focused on holding individuals accountable can often be manipulated. When they are, they report inaccurate information on the state of the system they monitor. Diagnostic metrics that are explicitly used only to assess the system and not used to hold individuals accountable for the performance of the system are typically more accurate and reliable. Any complex system typically benefits from using both types of metrics in tandem—motivational metrics to induce system performance and diagnostic metrics to ensure that the system being monitored is actually doing what the motivational metrics are reporting. In principle, an Air Force compliance inspection could use both types of metrics in this way. But voluntary reports of the kind described in the text can typically only support diagnostic metrics reliably.

to ensure that reporting is voluntary, confidential, and non-punitive.[4] ASRS's current operation is relatively modest, with an average annual budget of about $2.4 million, and 35 employees,[5] which includes four full-time IT specialists.

The *Aviation Safety Action Program*, like ASRS, seeks to encourage voluntary reporting of safety information that may be critical to identifying potential precursors to accidents. ASAP works at the operator level and targets the employees of FAA certificate holders. The program has industry-wide participation.

Event Review Committees (ERCs) transform safety issues raised by voluntary reporting into corrective actions. An ERC includes three representatives, one each from the FAA, the certificate holder, and a third party, typically the employee group's labor union. Each carrier or operator has a set of ERCs, one for each employee type—for example, pilots, flight crew, mechanics, and dispatchers.

ERCs play a role similar to that of SMEs in the ASRS. They review safety incident reports, perform triage, and, based on the current guidelines, decide on a course of action within 24 hours. The ERCs regularly compile lessons learned with emphasis on improving the safety culture of the FAA, certificate holder, and the labor union. Lessons learned cover areas like training, ERC processes and teamwork, and ASAP data collection/analysis/dissemination. ERCs share these lessons across the aviation community.

ASAP uses a commercial off-the-shelf (COTS) database reporting system. The FAA pays for the COTS system, but other expenses, including the personnel costs associated with the ERCs, are borne by the certificate holders and other participating organizations, such as labor unions.

The Voluntary Disclosure Reporting Program is an operator-level program that provides positive incentives to FAA certificate holders who voluntarily identify, report, and correct their own instances of regulatory noncompliance. When a certificate holder discloses and immediately corrects an unintended regulatory violation, the FAA takes lesser enforcement (often administrative) actions with the certificate holder.

For a report to be covered under the VDRP, the current guidelines require that a certificate holder meet the following conditions:

- The certificate holder notifies the FAA immediately after detecting apparent violations (initial notification within 24 hours, followed by a written report submitted to the FAA within 10 working days of initial notification).
- The apparent violation (a) was inadvertent and (b) does not indicate that there was a lack (or a reasonable question) of qualification.
- Immediate action was taken upon discovery to terminate the mishap conduct.
- The certificate holder has developed, or is developing, a comprehensive fix and a schedule of implementation satisfactory to the FAA.

The FAA works closely with the certificate holder in monitoring the implementation of the fix. To close a VDRP infraction case, the FAA signs off when the implementation is complete.

---

[4]   NASA has been also supporting ASRS-type applications in other industries, including the healthcare, railroad, and maritime industries.

[5]   This includes both part- and full-time employees.

Flight Operational Quality Assurance is an operator-level system that collects digital data generated during normal in-flight operations. Unlike the ASAP and VDRP, which manage largely subjective data provided by volunteers, FOQA manages objective data recorded in flight while operational activities take place. FOQA data are often used to supplement the data received through the ASAP, VDRP, or other subjective voluntary programs. Any infractions discovered through FOQA are reported through the VDRP.

Findings from FOQA data also flow to the FAA's pilot training program, the Advanced Quality Program (AQP). AQP is an FAA safety training program geared toward individual airline pilots. AQP can replace airlines' own safety training programs for their pilots to satisfy the FAA's airman certification requirements. The program can be individually tailored to address specific safety issues for the individual pilots being trained. The AQP program provides a direct mechanism for the FAA to follow through with each pilot on any safety issues indicated by the FOQA data.

Voluntary Programs within Certificate Holders. Air carriers and certificate holders independently implement voluntary programs as part of their own internal evaluation systems. The Line-Oriented Safety Auditing (LOSA) program is an example of such a program. Under LOSA, an airline hires a third-party observer to monitor various aspects of its operations. For example, the third party might observe cockpit operations, including procedures, workarounds, tribal knowledge, etc., to catch what may or may not show up in the ASAP and FOQA. The infractions identified during these internal evaluations can also be reported through the VDRP.

Air Transportation Oversight System. The voluntary reporting programs described above feed into the ATOS, a comprehensive database system that FAA PIs and managers use in the certification[6] of air carriers.[7] The ATOS supports the design assessment, performance assessment, and risk management processes described in Chapter Two.[8] PIs currently use the ATOS as their primary tool in the inspection and certification processes for all air carriers. This system helps the FAA identify safety hazards and risks that formal inspections cannot identify or anticipate, in particular those not reflected in current regulations. This helps the FAA assess and anticipate future regulatory needs. Air carriers also voluntarily share their proprietary and confidential information through a database system called the Aviation Safety Analysis Information Sharing (ASAIS) program, which provides input to the ATOS certification process.[9]

---

[6] Most of the FAA inspections are performed as part of the certification of air carriers and other commercial aviation operators, and, therefore, the terms *inspection* and *certification* are loosely and interchangeably used in this report.

[7] The building blocks of FAA inspections are Certificate Management Offices (CMOs), Certificate Management Teams (CMTs), PIs, and individual staff inspectors. In general, the FAA establishes a CMO for each carrier and assigns a CMT. A CMT, administered through a CMO, provides continuity for the carrier through permanent members that are assigned to and trained on the specific procedures of the carrier in question. A CMT generally consists of (1) PIs in each major functional area (i.e., maintenance, operations, avionics, etc.), (2) groups of staff inspectors that support each PI, and (3) other administrative and specialty staff that support the management and operations of the CMO. The functional responsibilities and expertise of FAA inspection and certification activities lie with the PIs.

[8] FAA Order 8900.1, *Flight Standards Information Management System (FSIMS)*.

[9] ASAIS functions as a repository for industry-wide voluntary reporting data that are rolled up at various levels of aggregation, such that industry-wide observations, trends, and other analyses can be conducted. One of the goals of ASAIS is to perform aggregate industry-level analyses so that the voluntary reporting programs can be used as a more predictive tool for industry safety trends.

Cost Effectiveness of the FAA Voluntary Reporting Programs

FAA voluntary reporting programs have had significant cost implications for the FAA and air carriers. Although the actual reporting is done by the airlines and their employees, the oversight and monitoring of these programs require FAA resources. It appears that airlines have been able to absorb their costs because the reporting programs affected all airlines in the United States and so did not impact the relative competitiveness of individual carriers. The programs may even have promoted performance improvements that, over time, offset the costs imposed by the new safety rules. Even without this effect, however, the air carriers were able to pass new costs on to their customers. According to the FAA, it was able to implement the programs without a significant budget increase. Most of the additional costs associated with the voluntary programs were absorbed by the FAA's inspection budget. This increase in costs was made up by a significant cutback in the routine surveillance activities[10] and by the efficiency gains achieved through the automation of inspection processes as part of the ATOS implementation.

According to the FAA, no one has formally analyzed the effectiveness of the voluntary reporting programs. In particular, no one has quantitatively tracked[11] performance metrics of the programs. Based on anecdotal observations by FAA PIs, management, and other personnel involved in safety, there is unanimous agreement that the voluntary reporting programs have been "well worth the effort," and that "the payback was very high." This consensus is based on the sheer number of incidents the programs reported, the significant safety issues these reports identified, and the subsequent corrective actions taken. There is a consensus that the surveillance activities in place prior to the voluntary programs would not have identified most of these safety issues and proactive corrective actions would not have been possible.[12]

Implications for the Air Force Inspection System

Each of the FAA voluntary reporting systems described above offers capabilities that the Air Force might consider instituting. The industry-wide ASRS points to the potential for an analogous Air Force–wide system that gathers information from outside the traditional compliance channels and assesses that information to gain new insights about how the Air Force works and how compliance standards applied in traditional inspections should change. Voluntary reporting outside of the official inspection channels allows the Air Force to get ahead of the curve and proactively adjust its policies to avoid problems the current compliance system cannot even detect.

A program similar to the operator-specific ASAP could target personnel within wings. As described below, the Air Force Safety Center currently runs such a system. Experience with this program in the AMC, the only Air Force organization using the program to date, has been positive.

---

[10] Routine surveillance here refers to those routine inspection activities that were no longer part of the specific inspection processes outlined in the ATOS system.

[11] David Gilliom from the FAA Flight Standards indicated that past attempts at quantitative and trend analyses were not all that useful. He found most useful information came from qualitative comments. The recently fielded ASAIS system mentioned earlier is a more formalized attempt by the FAA to perform industry-wide quantitative and trend analyses of the data received through the voluntary reporting programs.

[12] By itself, the FAA experience cannot tell us how voluntary reporting would perform in the context of an Air Force compliance inspection, but it points to the potential for substantial net benefits. Such benefits would accrue to the Air Force, of course, only once it had adapted voluntary reporting to its own cultural and organizational setting.

CHAPTER FIVE

Introducing the New Unit Effectiveness Inspection (UEI)

SAF/IG's vision for a new inspection system calls for the complete overhaul of the Air Force's current compliance inspections. At the time of this report, the plan was for compliance inspection to become a thing of the past; only certain elements will be preserved in a new type of inspection, the UEI. This IG-led inspection has two distinct elements. First, the inspection team will verify and validate the wing's CCIP (described in Chapter Four). This consists of inspecting special-emphasis items and a sample of core compliance areas using methods such as interviews, surveys, task evaluations, audits, and program reviews. The second component of the UEI is an assessment of the wing's discipline, leadership, and aspects of climate or culture. The intent of this assessment is to answer the "Big 7" questions pertaining to readiness, compliance, proficiency, and effectiveness.

This chapter focuses on the second element of the UEI. Specifically, we report findings from our fieldwork and summarize scholarly research from several fields as it relates to leadership, leadership climate, and discipline. This includes a definition of leadership and discipline, and an examination of their relationship with performance. Also, particular attention is paid to the measurement of leadership and discipline. We describe how Air Force personnel currently measure these factors or believe they can be measured; highlight ways that leadership, in particular, has been effectively measured by academics; and offer for consideration two measurement tools already used by the Air Force: the Air Force Manpower Agency's (AFMA's) Climate Survey and AFSC's AFCAST surveys. We conclude the chapter by noting the reservations some of our fieldwork participants had regarding the measurement of leadership within the context of inspection.

Leadership

Definition

General Wilbur L. "Bill" Creech, past commander of the Tactical Air Command, said, "There are no poor outfits, just poor leaders. . . . The leadership makes all the difference—always" (1994, p. 349). The topic of organizational leadership has been the subject of thousands of studies, primarily in the fields of psychology and business, over the past century (Kaiser, Hogan, and Craig, 2008). A recent review of such literature included over 1,100 peer-reviewed research articles published within the past 25 years (Hiller et al., 2011). Although it is so widely studied, little consensus exists on how to define leadership. This may be due, in part, to the diversity of topics that fall under the banner of leadership research. Kaiser and his colleagues suggest leadership studies fall into two research traditions: how leadership emerges and how effective it is

(2008). Leadership *emergence* pertains to the process of exercising influence over individuals in order to achieve status or another goal, and includes a large body of work related to leadership traits and leadership behaviors measured using a variety of methods. Leadership *effectiveness* studies also vary greatly; some studies focus on member satisfaction and morale, and others focus on more objective measures like productivity or attrition (Kaiser, Hogan, and Craig, 2008). Hiller and his colleagues suggest that scholars have sought to answer six major questions about leadership, albeit via largely independent, non-complementary streams of research (2011, p. 1139):

1.   From whose perspective is leadership judged?
2.   Which type of leadership measure is used (i.e., the method of collecting data)?
3.   On which criterion domains are leadership effects assessed (e.g., effectiveness, behavior, motivation, or cognitive)?
4.   At what time frame are leadership criteria being examined?
5.   At what level of analysis are leadership criteria being examined?
6.   What is the organizational level at which leadership effects on criteria are being examined?

The lack of a dominant paradigm in leadership research helps explain the absence of both a specific definition of leadership and a consensus on how to measure it. There appears to be general agreement, however, that leadership involves a process of influence in which a leader motivates his or her followers to work toward a specific goal. Simply put, leadership involves "bringing people together and combining their efforts to promote success" (Kaiser, Hogan, and Craig, 2008, p. 96) or, as General Creech put it, "creating common perspectives and common purpose" (1994, p. 380). We used this view as a starting point for our study and this chapter, but also delved into appropriate lines of leadership-related research, given SAF/IG's vision for the Air Force inspection system.

During our focus group sessions and interviews with IG inspectors, members of recently inspected wings, and the leadership element of those wings, we explored how leadership was defined within the Air Force context. We found, as with the scholarly literature, that leadership was an amorphous concept, and that some personnel found it hard to formulate a definition. The most frequently mentioned aspects or qualities of leadership included the following:

- proactive
- knowledgeable
- hands-on and involved, yet not too "in the weeds"
- able to instill pride and a sense of purpose
- able to motivate people, even to do "something they hate."

Examples of remarks made by Air Force personnel during our focus groups and interviews are provided in Table 5.1.

Relationship with Performance

Perhaps due to the variety of definitions offered by the personnel we spoke with and the range of leadership attributes emphasized, there was also a lack of consensus on whether leadership affects performance, compliance in particular. Among both inspectors and inspectees, there

Table 5.1
Evidence from Fieldwork on Defining Leadership

Leadership Attribute	Examples
Proactive	Leadership needs to be proactive and . . . optimistic, for lack of a better word (Inspector, focus group 2).
	Assertiveness and proactiveness (Inspectee, officer focus group 2).
Knowledgeable	Engagement, if [the leader] is knowledgeable enough that he can answer me, rather than calling his section chief. He should know something (Inspector, focus group 7).
	Good leaders will articulate how they set themselves apart. [Their leadership] will show in how their airmen express their program. They will be knowledgeable above and beyond the minimum AFI regulations (Inspector, focus group 2).
Integrity	Integrity, and if the person is respected personally and professionally (Inspectee, leadership interview 1).
Hands-on and involved	Are they involved? Maybe not in the weeds, but involved. If a self-inspection is done, and we arrive [and do our own inspection], and all the answers were red but leadership says, "My people said all is good to go," then probably they were not that involved (Inspector, focus group 6).
	Leaders need to be visible in the work center. They don't need to be experts, but they need to be out there so that airmen know they care (Inspectee, NCO focus group 2).
Able to instill pride, purpose	Leadership is the ability to motivate people, to explain why what they are doing is important (Inspectee, leadership interview 6).
	You can see it down to the lowest airmen. If he knows why he is important as far as getting the mission done, [leadership is good] (Inspector, focus group 9).
	Leaders instill a sense of pride so they go out and do things not because they're required (Inspector, focus group 2).
Able to motivate	The truest measure of an effective leader is the willingness of his subordinates to do something that they either hate doing, or that will land them in a world of hurt. If he [the leader] tells them [his subordinates] to mow the grass in 100-degree heat, his subordinates take a deep breath and go do it without complaining. It is also the truest measure of morale; doing something you hate because it's the right thing (Inspectee, leadership interview 12).
	Motivation really is big for us, not to mention there are many forms of leadership depending on the goals. We're spread thin, and we have lots asked of us. Motivating people to do all that's asked of them and then to be compliant on top takes leadership (Inspectee, leadership interview 2).

SOURCE: 2011 RAND fieldwork for SAF/IG.

were some who thought leadership had no relationship with compliance, others who thought the two were somewhat related, and still others who believed leadership strongly influenced compliance. Illustrative remarks for each of these opinions are provided in Table 5.2. Further, a few individuals noted that some units succeed in spite of poor leadership, while one study participant suggested that deficiencies in non-mission critical areas were indicative of good leadership:

> The fact that a unit has deficiencies in non-mission areas might actually indicate good leadership. If they take resources away from one area and put it elsewhere, as a priority . . . I don't think deficiencies or compliance predicts leadership (Inspector, focus group 4).

Table 5.2
Evidence from Fieldwork on the Leadership-Performance Link

Link Type	Examples
No relationship between leadership and compliance	No, there's no direct correlation. It's not indicative of compliance. There's also a difference between compliance and mission accomplishment. You can be not in compliance, but still getting the mission done, perhaps well (Inspector, focus group 4).
	I thought that this [interview] was about compliance? [He didn't understand the discussion of leadership in relation to compliance]. I would say that the UCI is an exercise in management, not leadership (Inspectee, leadership interview 6).
	I don't think it all goes back to the commander. Ultimately, the commander is responsible for everything, but there are so many gears. There are so many duties a commander has to fill. Some of those gears are telling a commander that the program is good to go, and then we some and do the compliance inspection, and we find lots of problems. They are told they are fine on the self-inspection, but we find that they aren't. There are too many pieces of the puzzle for the commander to fit together (Inspector, focus group 4).
Some relationship between leadership and compliance	You can be a great leader but not compliant, and you can be 100-percent compliant, working compliance all day, and be a poor leader. I'm not sure the two are completely interconnected. There is some relationship [between the two] but I'm not sure of exactly what it is (Inspectee, leadership interview 8).
	Leadership will amplify the results that are in the direction they were going anyway. Strong leadership will achieve the best possible outcome. With weak leadership, there will be drawbacks [in compliance ratings], but it [weak leadership] won't throw you entirely off the track. The Air Force is designed to overcome sub-optimum traits in leadership (Inspectee leadership interview 12).
	If there's a bad grade, there is likely bad leadership (Inspector, focus group 5).
Strong relationship between leadership and compliance	Participant #1: It's [leadership] critical to doing well because it will ensure we are mission ready and inspection ready all the time. Bad leadership allows you to develop bad habits that you spend months correcting. The good thing about any inspection is it gets you [unit] back to the baseline. Good leadership keeps you close to that mark, and bad leadership. . . .
	Participant #2: It allows you to drift. He's right. Six to eight weeks outside the inspection, you're still doing good. And depending on leadership, you start to deviate. You start to deviate and then have to re-center yourself for the next inspection (Inspectee, NCO focus group 2).
	Leadership is so important in compliance because if the colonel tells me to do something because of compliance I will do it, less so if it is not important to him. There will be compliance, but not enthusiastic compliance, if the commander says it [the particular compliance item] is really stupid even though he said to do it. Leadership absolutely affects compliance (Inspectee, leadership interview 4).

SOURCE: 2011 RAND fieldwork for SAF/IG.

Turning our attention to the literature on leadership, scholars have debated the relative impact of leadership versus that of other factors on performance. Some assert that leadership's importance is overemphasized due to "leader attribution error," a tendency to identify a leader as the primary influence on collective performance (Hackman and Wageman, 2007). Whether leadership matters—i.e., whether it influences performance—depends both on the level of leadership under consideration and on how leadership and performance are measured. Our review of leadership research suggests it *does* matter, based on two types of research in particular: studies of top leadership and studies pertaining to transformational and transactional leadership.

Studies of top leadership focus on how changes in leadership at the top level, such as a chief executive officer (CEO) or mayor, are related to changes in organizational performance.

These studies often employ a statistical technique called "variance decomposition," which essentially uses information about an organization (e.g., size, revenue), the industry in which it operates (e.g., manufacturing, services), and the context (e.g., year) to determine what changes in performance, often measured as return on assets, can be explained by the change in CEO-level leadership. Overall, these studies (e.g., Lieberson and O'Connor, 1972; Mackey, 2008; Salancik and Pfeffer, 1977; Weiner and Mahoney, 1981) demonstrate that, while leadership typically has less of an effect than do characteristics of the organization, it still accounts for a notable portion of changes in performance. As Mackey notes, there have been criticisms of this type of approach, particularly related to the statistical methods used, but refinements have yielded supporting results (2008).

Another shortcoming of this line of inquiry is that it does not take into consideration the inner workings on an organization or what a leader actually does to influence the behavior of others in pursuit of an objective. However, the robust literature on two types of leadership, transformational and transactional, does provide such insights and has found that leadership has significant effects on performance. Initially described by Bass in his seminal work, *Leadership and Performance Beyond Expectations*, transformational and transactional leadership have been studied by numerous researchers for more than 25 years (1985). Transformational leadership occurs when a leader tries to increase followers' awareness of what is right and important and motivate them to perform at higher levels, while transactional leadership is more concerned with clarifying goals and rewarding behaviors that are sufficient within an existing system. Dvir and his colleagues distinguish between the two as follows (2002, p. 735):

> *Transactional leaders* exert influence by setting goals, clarifying desired outcomes, providing feedback, and exchanging rewards for accomplishments. *Transformational leaders* exert additional influence by broadening and elevating followers' goals and providing them with confidence to perform beyond the expectations specified in the implicit or explicit exchange agreement.

While Zohar explains the distinction as such (2002, p. 88):

> The (transactional) supervisory role has to do with the organization of tasks and with getting people to do things more reliably and efficiently, whereas the (transformational) leadership role has to do with development and with getting people to commit themselves to more challenging goals.

A number of studies have documented a positive relationship between both types of leadership and performance (see Wang et al., 2011, for a review). This includes several studies based in a military setting. For example, Dvir et al. conducted a field experiment with Israel Defense Forces personnel and identified a link between transformational leadership training and cadets' performance six months after that leadership training took place (2002). Bass and his fellow researchers studied U.S. Army platoons conducting a combat simulation exercise and found that the ratings of both transformational and transactional leadership exhibited by the platoon leaders were linked with platoon performance in the exercise (2003). Similarly, Curphy demonstrated that the transactional and transformational leadership of Air Force squadron leaders were related to squadron motivation, cohesion, and performance (1992).

The relative impact of transactional and transformational leadership has varied depending on the type of performance measured, such as individual task performance versus overall unit performance (Wang et al., 2011). Further, Bass suggests that the type of leadership necessary may depend on the context, with transactional leadership more likely to be found in well-ordered conditions and transformational leadership employed in more dynamic settings (1985). Based on their Army research, Bass and his colleagues suggest that transactional leadership may be more important in a military context than in other settings, given both the need to execute many complex procedures and the typical rapid turnover of personnel. Although the conditions in which transactional or transformational leadership—or both—matter are still being explored, findings that both types of leadership influence performance are quite robust (2003).

Discipline

Definition

The term *discipline* is ubiquitous in the Air Force lexicon, yet we could not find an official definition of this term in the current set of Air Force publications. Air Force Policy Directive 36-29, *Military Standards* (2009, p. 2), suggests that discipline is multi-faceted:

> When wearing the uniform, all Air Force members will adhere to standards of neatness, cleanliness, safety, and military image to provide the appearance of a disciplined Service member.

Seemingly consistent with this view, when asked to describe what discipline is, how to measure it, and how it may affect compliance, Air Force personnel in our interviews and focus groups often struggled to do so. As one leader told us during an interview, "Measuring discipline is tough. Definitions are different everywhere" (Inspectee, leadership interview 7). Another asked, "Do you mean as a noun or verb?" (Inspectee, leadership interview 8). A number of times we received vague responses like, "You know it when you see it."

However, we also heard during these interviews and focus groups very specific ways in which Air Force personnel perceive discipline. Some viewed discipline in terms of adherence to technical specifications or following rules. Inspectors in particular discussed discipline in terms of customs and courtesy. Still others offered definitions of discipline rooted in legal codes and infractions. Finally, discipline was occasionally mentioned in terms of meeting physical fitness standards. Table 5.3 provides examples of the different definitions of discipline given during our interviews and focus groups with Air Force personnel.

Scholarly literature does not appear to offer a resolution to the lack of consensus on how discipline is defined. In stark contrast to leadership, there is a dearth of research on discipline in military contexts. In some ways, transactional leadership, with its emphasis on rewarding satisfactory adherence to codified procedures, seems related to discipline, yet is not a perfectly interchangeable concept.

Relationship with Performance

As was the case with our discussion of leadership, during interviews and focus groups, Air Force personnel differed in their opinions of whether and to what extent discipline influences performance, specifically in terms of compliance. Most who offered an opinion felt that discipline likely affected compliance to some degree. Remarks of this nature include the following:

Table 5.3
Evidence from Fieldwork on Defining Discipline

Discipline Type	Examples
Adherence to orders, following rules	Knowing what's required and adhering to it, willfully (Inspectee, leadership interview 4).
	A unit doing the right thing without being watched (Inspector, focus group 9).
	If a unit is disciplined, they are adhering to the rules (Inspector, focus group 5).
	If disciplined, then that means they are consistently doing what it is they're supposed to be doing (Inspector, focus group 5).
	Operational discipline is doing the task as spelled out in the book (Inspectee, leadership interview 7).
Customs and courtesy	If they stand up to meet and greet when you come in, offer you coffee and donuts, they are respectful. You can definitely tell (Inspector, focus group 8).
	It is the whole package. You are walking in the door. Does the unit take pride in the facility and place they live? If the place looks like trash, it gives me an impression that [the unit] doesn't care. But, an old facility doesn't mean [the unit is] trash. Do people take pride in their uniforms, how they talk, are they taking that time [to be proper in their speech] or do they not?" (Inspector, focus group 9).
	Customs and courtesy . . . [p]eople at attention when O-6s, NCOs, walk into the building . . . the basics, the small things (Inspector, focus group 3).
	If there's no discipline, when an officer comes into the room, no one stands up (Inspector, focus group 1).
Legal codes and infractions	I look at UIFs [Unfavorable Information Files]. Sometimes someone may not get a DUI, but something goes into a UIF. Did that person get the same punishment as someone else who did the same thing? Article 15s, DUIs . . . we look at them. That's how I measure it (Inspector, focus group 6).
	There is a good order and discipline in the military. One that we use as a measure is DUI rates, Article 15s, non-judicial punishments, fitness failures, court martials . . . that goes into good order and discipline in squadron (Inspectee, leadership interview 1).
	DUI rates, assaults, non-judicial punishment, court martials, how we meet EPR requirements, are we launching recovery aircraft like we are supposed to? (Inspectee, leadership interview 4).
	Leadership and discipline are not part of the UCI, but we have "state of discipline" meetings quarterly where we go across all of the discipline reports and make sure we are hitting the bell curve in terms of reprimand to offense. We check to ensure that we have similar responses to similar offenses (Inspectee, leadership interview 5).
Physical fitness standards	For example, with fitness, we are almost at a 100-percent on-time with compliance, with a 96-percent passing rate. Those are things I think we should look at for discipline (Inspectee, leadership interview 4).

SOURCE: 2011 RAND Fieldwork for SAF/IG.

I don't think you can write reports that are bad without bad leadership and discipline (Inspector, focus group 9).

Discipline leads to compliance—I think that's true. Leadership is critical, but there are a lot of things that are critical. A disciplined pilot will be a good pilot, but also good on more subjective things. Compliance is a positive outcome of good discipline (Inspectee, leadership interview 2).

[If a unit is lax with customs and courtesy], they probably won't be up on their programs either. One gauge [of how well a program is likely doing] is when you are talking to someone and they answer three to five questions for you in advance, before you ask the question. Those people want to tell you about their program, and explain to you how it works. If the person just says "yes," gives you one-word answers, then you will probably have to pull all the information out of them. . . . Units that are disciplined will give you answers, will even identify shortcomings but will also go on about their plans to correct them, how they are going to bring things up in that area versus forcing us [inspection team] to find it (Inspector, focus group 2).

Again, in resemblance to the comments about leadership, the perception existed that a highly disciplined unit may deliberately not be in compliance with the standards regulating less critical areas such as commander's programs.

Measurement of Leadership and Discipline

Approaches Suggested by the Literature
Transactional and Transformational Leadership. Earlier in this chapter, we discussed the large body of research on transactional and transformational leadership and noted that evidence of the relationship between both types of leadership and performance was quite robust. Much of this research was conducted using the Multifactor Leadership Questionnaire (MLQ). The current version of the MLQ is a proprietary 45-item questionnaire developed in 1997 by two psychologists, Bass and Avolio. It measures nine distinct leadership factors and three leadership outcomes, and has been both extensively validated (e.g., Muenjohn and Armstrong, 2008) and used by scholars, consultants, and other practitioners in a wide variety of settings worldwide. Among the leadership factors measured are three dimensions of transactional leadership and four dimensions of transformational leadership, listed in Table 5.4.

The MLQ is relatively easy to administer and takes about 15 minutes to complete. In preparation for an inspection visit, the IG could have a random sample of wing personnel complete the MLQ with regard to a specific leader. These perspectives of subordinates, peers, and superiors could provide a more comprehensive measure of the leader's effectiveness. Taking this premise further, MLQ-based ratings of multiple leaders at different levels of hierarchy, both NCOs and officers, could be collected and analyzed to assess the effectiveness of the wing's full leadership chain. Just as a CEO may not be entirely responsible for the success or failure of his organization, so too may a wing commander not deserve full credit—or blame—for his or her unit's functioning. A statistical technique called "data aggregation" (see O'Reilly et al., 2010, for an example) could be used to determine the combined effect of leadership across the wing, from flight up through wing, from superintendent to wing commander.

Given the lack of research on discipline in the military setting, it is possible that measures of transactional leadership collected in this manner could be combined with other measures of discipline, like customs and courtesy or physical fitness ratings, as well as with measures of the *lack* of discipline, such as instances of driving under the influence and non-judicial punishments, to obtain a multi-faceted measure of discipline.

Psychological Safety. In our earlier discussion of self-reporting practices (Chapter Four), we introduced the concept of psychological safety and discussed how leadership can influence

Table 5.4
Dimensions of Leadership Measured in MLQ

Transactional Leadership
Contingent reward: Rewards for good performance, sanctions for bad
Management-by-exception (active): Monitors to detect mistakes
Management-by-exception (passive): Intervenes after problems arise
Transformational Leadership
Idealized influence: Serves as a role model with clear vision and a sense of purpose
Intellectual stimulation: Challenges followers to solve problems
Individualized consideration: Provides teaching and coaching to followers

SOURCE: Judge and Piccolo, 2004.

whether an organization's members believe it is a safe place for risk taking, to include the voluntary disclosure of errors. Accordingly, how successful a leader is in fostering an atmosphere of psychological safety may be a particularly useful measure in the context of a UEI. It may hint not only at the potential for learning within a unit but also at the likelihood that proactive self-reporting of deficiencies occurs. As Edmondson suggests in her study on error reporting in a hospital setting, when leaders establish an atmosphere of openness in discussing mistakes, it is likely to affect their rate of disclosure (1996). In other work, Edmondson found that team coaching by a leader was related to perceptions of psychological safety, which in turn influenced learning behaviors and performance (1999). The published version of this study included survey items for both psychological safety and leader coaching that have been used by other researchers. Shown in Table 5.5, these Likert-scale items could be adapted to the Air Force context and employed by IG personnel to gauge the level of psychological safety within a wing.

Moreover, as with transactional and transformational leadership, data could be obtained about leaders at different levels and then aggregated to gain a collective measure of psychological safety throughout the wing.

Approaches Suggested by Current Air Force Practice

Air Force Climate Survey. A large-scale survey administered by the Air Force for a number of years, the Air Force Climate Survey, may also be of use within the context of the new UEI. AFMA fields this survey throughout the Air Force every two years (Salomon, 2003). The most recent survey was fielded in October–November 2010 and yielded 173,000 responses from about 600,000 individuals queried (AFMA, undated). Individuals completed a web survey that included questions about unit "inputs (things about the job, unit-level resources, and core values), organizational processes (supervision, leadership, training and development, teamwork, recognition, and unit flexibility), and outcomes." These responses provided the basis for a "system-wide analysis of [each] unit's organizational climate" (AFMA, 2002). When AFMA reports survey results, it carefully masks those results in which there are so few respondents that it might be possible to identify individuals' responses via inference.

The survey can generate reports for any squadron-equivalent unit with at least 20 respondents. It can generate detail on the responses of officers, enlisted personnel, government civilians, and Non Appropriated Fund personnel whenever there are responses from at least seven

Table 5.5
Edmondson's Measures of Psychological Safety and Team Leader Coaching

Psychological Safety
If you make a mistake on this team, it is often held against you.
Members of this team are able to bring up problems and tough issues.
People on this team sometimes reject others for being different.
It is safe to take a risk on this team.
It is difficult to ask other members of this team for help.
No one on this team would deliberately act in a way that undermines my efforts.
Working with members of this team, my unique skills are valued and utilized.

Team Leader Coaching
The team leader . . .
. . . initiates meetings to discuss the team's progress.
. . . is available for consultation on problems.
. . . is an ongoing "presence" in this team—someone who is readily available.

SOURCE: Edmondson, 1999.

people in a category. (Reporting data with fewer respondents could compromise the survey's confidential nature.) This generates extensive data on a wing's leadership and potentially offers fine detail within a wing. While administering the survey every other year is compatible with the proposal to hold a major inspection for every wing every other year, the data collected might cover a period when the previous command staff was in command of a wing. Even if the IG did not incorporate findings from the survey into its own inspection planning and execution, it could use the definitions of leadership offered in the survey in its own setting. It could even field its own survey during each major wing inspection, perhaps one that is shorter and more focused, to reduce any burden on the inspected unit. The next section illustrates a survey approach that requires fewer questions.

Of particular relevance here, the Climate Survey collects extensive information on unit leadership. Thirty-nine of the 98 questions on one recent survey directly addressed leadership and many of the remaining questions yielded information still relevant in some way to measuring the quality of leadership.[1] The questions specifically targeting leadership ask about it at three different levels:

- perceptions of the behavior of a respondent's direct supervisor
- perceptions of the behavior of the commander of a respondent's unit
- perceptions of the chain of command as a whole within a respondent's unit.

---

[1]   2003 Air Force Climate Survey. The quotations that follow come from this survey. The language we use to describe each point is drawn directly from the language in the survey. Appendix D provides a list of the remaining questions in the survey.

Table 5.6
Air Force Climate Survey Measures of Direct Supervisor

My direct supervisor . . .
. . . is good at planning my work.
. . . sets high performance standards.
. . . is concerned with my development.
. . . corrects poor performers in my work group.
. . . looks out for the best interests of my work group.
. . . provides instructions that help me meet his/her expectations.
. . . helps me understand how my job contributes to my unit's mission.
. . . ensures that there is a fair distribution of the workload among the people.
. . . provides opportunities for me to give feedback to him/her.

SOURCE: 2003 Air Force Climate Survey.

Questions about the direct supervisor include: "[h]ow you are being utilized, organized, led, and provided feedback?" Respondents use a six-point Likert scale to indicate the extent to which they agree or disagree with a series of statements about their direct supervisor, listed in Table 5.6.[2]

The Climate Survey seeks two kinds of information about unit commanders. It first asks each respondent to comment on the "practices and behaviors of your unit commander, commander equivalent, or director." The survey assures respondents that the data collected "will not be used in any form for performance evaluations." Respondents again use a six-point Likert scale to indicate the extent to which they agree or disagree with the statements provided in Table 5.7. Several of the practices and behaviors referred to in these statements, such as the ability to motivate personnel, were mentioned during our fieldwork as important leadership practices.

Another set of questions on the Climate Survey focuses on a different category of behaviors exhibited by a unit commander or commander equivalent. Specifically, it asks respondents to note "how frequently you feel your unit commander (or commander equivalent) exhibits the behaviors described." These behaviors, described in Table 5.8, resemble individual leadership traits and again map onto the attributes cited by focus group and interview participants as key elements of leadership—such as being knowledgeable, possessing high integrity, and caring for the unit. For these measures, a five-point Likert scale is employed to structure the answers.[3]

Finally, the survey addresses the chain of command as a whole. In this manner, the survey is reminiscent not of our fieldwork, but rather of scholarly work on leadership, namely, the practice of using data aggregation to assess the effect of the entire leadership chain. Specifically, the survey instructs respondents to consider how the "chain of command in your unit is influencing the direction, people, and culture of the unit. This includes all levels from your

---

[2] Responses allowed are "strongly disagree," "disagree," "slightly disagree," "slightly agree," "agree," "strongly agree," and "don't know."

[3] Options are "never," "seldom," "sometimes," "generally," "always," and "don't know."

Table 5.7
Air Force Climate Survey Measures of Unit Commander or Commander Equivalent

My unit commander (or commander equivalent) . . .
. . . sets challenging unit goals.
. . . provides a clear unit vision.
. . . makes us proud to be associated with him/her.
. . . is consistent in his/her words and actions.
. . . is inspirational (promotes esprit de corps).
. . . motivates us to achieve our goals.
. . . is passionate about our mission.
. . . challenges us to solve problems on our own.
. . . encourages us to find new ways of doing business.
. . . asks us to think through problems before we act.
. . . encourages us to find innovative approaches to problems.
. . . listens to our ideas.
. . . treats us with respect.
. . . is concerned about our personal welfare.

SOURCE: 2003 Air Force Climate Survey.

supervisor to your unit commander, commander equivalent, or director." The survey uses a six-point Likert scale to structure responses to the items listed in Table 5.9.

All in all, these questions concretely operationalize many of the ideas about leadership discussed earlier in this chapter. They effectively articulate a current official Air Force definition of what good leadership is, even if the survey makes no judgments (and asks for none) about the relative importance of the attributes measured. In effect, the survey yields a scorecard for a unit's leadership that Air Force leaders and analysts can use by imposing their own judgments of relative importance.

Air Force Culture Assessment Safety Tool (AFCAST). AFCAST is a family of closely related web-based surveys that senior leaders and unit commanders can use to assess their members' perceptions of safety issues and other operational factors relating to safety. Currently, AFSC provides 11 different AFCAST online surveys. Four surveys cover operations, maintenance, support, and higher headquarters activities in units with only non-nuclear systems. Five cover operations, maintenance, medical, and two kinds of support activities in units with nuclear systems. A more specialized survey covers private vehicle safety. The Voluntary Protection Program survey is designed to address broader federal policy promulgated by OSHA.

During interviews with personnel familiar with AFCAST, we learned that unit commanders can ask AFSC to field any of these surveys for activities within their units. Squadrons with nuclear systems are required to use it. Use elsewhere varies dramatically across MAJCOMs and across the individual surveys. The driving safety survey has been especially popular because squadron commanders resent losing the capabilities of hard-to-recruit, skilled

Table 5.8
Air Force Climate Survey Additional Measures of Unit Commander or Commander Equivalent

Measure	Definition
Integrity	Consistently adheres to a moral or ethical code or standard and considers the "right thing" when faced with alternate choices.
Organizational loyalty	Is devoted and committed to the organization.
Employee loyalty	Is devoted and committed to co-workers and subordinates.
Selflessness	Is genuinely concerned about the welfare of others and willing to sacrifice one's personal interest for others and the organization.
Compassion	Shows concern for the suffering or welfare of others and provides aid, or shows mercy for others.
Competency	Is capable of executing responsibilities assigned in a superior fashion and excels in all task assignments. Is effective and efficient.
Respectfulness	Shows esteem, consideration, and appreciation of other people.
Fairness	Treats people in an equitable, impartial, and just manner.
Self-discipline	Can be depended upon to make rational and logical decisions (in the interest of the unit).
Cooperativeness	Is willing to work or act together with others in accomplishing a task or some common end or purpose.
Sociability	Acts in an enthusiastic, friendly, and courteous manner towards others. Communicates in tactful and diplomatic ways. Provides a positive atmosphere.

SOURCE: 2003 Air Force Climate Survey.

personnel to non-mission-related accidents. Squadrons request the operations and maintenance surveys more than the support surveys.

When a survey window opens for an individual unit, personnel have 30 days to fill out the survey online, which typically has about 60 questions that use a Likert scale to categorize answers.[4] Additional questions allow free-text responses if personnel would like to provide more detail, particularly about *why* they chose the answers they did to the questions allowing

Table 5.9
Air Force Climate Survey Additional Measures of the Unit's Chain of Command

Measure
The leaders in my chain of command (in my unit) listen to my ideas.
The leaders in my chain of command (in my unit) are easily accessible.
I trust the leaders in my chain of command (in my unit).
I am proud to be associated with the leaders in my chain of command (in my unit).
I see the leaders in my chain of command (in my unit) doing the same things they publicly promote (walking the talk) and leading by example.

SOURCE: 2003 Air Force Climate Survey.

---

[4] AFCAST uses a five-point Likert scale that offers answers like "strongly agree," "agree," "neutral," "disagree," "strongly disagree," "not applicable," and "don't know."

only a Likert-scale response. Responses are anonymous and AFSC revises free-text answers to remove information that could violate anonymity before compiling reports. Participation is voluntary and no record is kept of who participated, so only the participants themselves know whether they even submitted answers. The unit commander controls access to information from the survey. Higher headquarters gain unrestricted access to survey information only after the information from enough organizations has been aggregated to prevent the higher headquarters from determing the performance of any one unit. AFCAST uses this anonymity to help ensure honest answers from respondents.

The surveys include questions about four different kinds of issues relevant to safety in a unit: organizational processes, organizational climate, resources, and supervision.[5] Table 5.10 displays the questions that the non-nuclear AFCAST surveys ask about supervision. The four columns on the right show the number of each question in each of the operations (Ops), maintenance (Mx), support (Sup), and higher headquarters (HHQ) surveys, respectively. If a cell in any of these columns has a number in it, a close analog of the question in the same row appears in the survey.

The IG might ask very similar questions in its own version of such a survey, but with a different emphasis. Rather than focusing on safety, the IG could highlight readiness, compliance, effectiveness, leadership, and discipline in a unit. Based on our assessment, questions shown in blue would require no change at all. Those shown in green could apply to a broader IG scope with only minor changes. The question in orange would require more substantial editing in order to pertain to the IG's mission. Only the questions in red would be difficult for the IG to adapt. To understand the ease of reframing such questions, note that a hazard, adverse incident, or human error need not have anything to do with physical safety. Risk management and quality assurance can be effectively applied to temper a broad range of potential failures that have nothing to do with safety per se.

Alternatively, the IG could take advantage of the AFCAST surveys as they currently stand. The justification might be to anticipate that a unit with problems related to safety is likely to have similar problems elsewhere. The AFIA has been considering the relationship between unit performance, as the IG measures it, and unit leadership, as AFCAST measures it.

This use of AFCAST presents three different kinds of challenges. First, a unit commander controls (1) when AFCAST surveys are applied to a unit and (2) the distribution of the survey findings. The IG could take advantage of an AFCAST survey only if it already exists and the commander is willing to share it. Assuring reliable IG access to AFCAST surveys would likely require a basic change in Air Force policy on AFCAST.

Second, such a change could easily compromise the validity of AFCAST survey findings. Currently, AFCAST carefully safeguards its survey data to encourage unit personnel to be candid in their responses. If a unit's personnel knew the IG would see their answers, they might be less forthcoming. Further, their leaders might be less aggressive about promoting participation and honesty. As we will report in the next chapter, we heard such concerns from Air Force personnel.

Finally, answers to questions framed in the context of safety concerns would be harder to interpret in an IG setting than answers to questions framed in broader terms. AFCAST measures perceptions. Psychologists have found that how respondents access these perceptions

5 Appendix E summarizes the full contents of the four non-nuclear surveys.

Table 5.10
AFCAST Questions About Supervision

Question, to Be Answered in a Likert Scale	Ops	Mx	Sup	HHQ
Leaders/Supervisors in my squadron are actively engaged in the safety program and management of safety matters.	46	48	42	36
Leaders/Supervisors in my squadron balance safety concerns with achieving mission tasking.		49	43	
Leaders/Supervisors are more concerned with operational tasks than safety.				37
Leaders/Supervisors encourage reporting safety discrepancies without fear of negative repercussions.	47	50	44	38
Leaders/Supervisors in my squadron set a good example for compliance with policies, rules, and instructions.	48	51	45	39
Leaders/Supervisors in my squadron permit cutting corners to get a job done.	49	52	46	40
Leaders/Supervisors in my squadron react well to unexpected changes.	50	53	47	41
Leaders/Supervisors in my squadron care for members' quality of life.	51	54	48	42
The Flight Safety Officer (FSO)/Missile/Space Safety Officer [squadron Safety Office/Safety NCOs/Safety Representatives/Safety personnel] is effective at promoting safety in my squadron.	52	55	49	44
Leaders/Supervisors in my squadron are successful in communicating safety goals to unit personnel.	53	56	50	
Leaders micromanage routine operations.	54	57	51	43
Operations Control Centers (e.g., Mission Operations Center [MOC], vehicle dispatch, Munitions Squadron [MUNS] Control, Security Control, etc.) are effective in managing work actions for my squadron.		58		
Work center supervisors coordinate their actions in my squadron.		59		
Contractors are held to the same safety performance standards as military and civilian Air Force employees.		60	52	

SOURCE: Air Force Safety Center, undated.

within themselves depends on how the questions are framed. An airman will think very differently about something as simple as who the relevant leadership personnel are when asked about safety as opposed to, say, readiness or effectiveness.

Users of data on the safety climate in an organization, like those provided by AFCAST, have warned us that the data must be interpreted with caution. Differences across wings or changes over time within a wing can result as much from changes in expectations about safety as from changes in the actual level of safety or the commitment to safety goals in a unit. This difficulty limits the usefulness of such surveys in inspections without instructions inspectors can use to interpret the survey results.[6] Understanding how and why perceptions differ from objective truth or reality will likely prove even more challenging if the Air Force uses perceptions about safety to make judgments about the broader realities relevant to the IG.

---

[6]   Users of AFCAST and related instruments seek additional information to put AFCAST results in context. In root cause analysis, AFCAST is best used as a place to begin looking for underlying problems, not as an immediate guide to what those problems are or how to mitigate them.

A third way the IG can use the AFCAST surveys is as a source of ideas for measures of leadership. In our interviews, we found general agreement that the effect of general unit leadership on safety cannot meaningfully be separated from its effect on any broader definition of performance. General unit leadership affects all the attributes of performance jointly. From the questions listed in Table 5.10, we infer that AFCAST detects an improvement in supervision (i.e., leadership) when personnel believe their unit leadership

- is actively engaged in the unit's operations
- encourages honest reporting of problems
- sets a good example for compliance with standards
- reacts well to unexpected problems
- cares about unit members' quality of life
- promotes efforts that are consistent with the unit's core goals
- communicates those goals clearly to unit personnel
- delegates authority effectively
- manages work actions effectively
- coordinates effort across work centers
- holds contractors to the same standards as unit personnel.

These points correlate well with the attributes of leadership discussed during our field-work and measured in some of the scholarly works we reviewed. AFCAST provides evidence that the Air Force values these specific attributes in a safety setting, and they could, therefore, offer a useful starting point for identifying what factors to emphasize in any IG effort to measure leadership in a broader setting, whether it uses a survey or not.

Concerns Regarding the Assessment of Leadership

Although our review of literature and Air Force practices suggests there are several ways to study leadership that have merit, during focus groups, inspectors aired some misgivings regarding the assessment of leadership in general, and within the context of a compliance inspection in particular. We note their concerns here because they represent potential psychological barriers that must be overcome before the new UEI concept can succeed. As the following comments demonstrate, some of the inspectors we spoke with feel that leadership is a subjective or emotionally charged concept that does not lend itself to checklist-based measurement:

> You can get an intuitive feel for the working environment, but it is difficult to apply a grade to that. That is very subjective and oftentimes very emotional. We have all worked for people who are nice but ineffective and also hard-asses who get the job done but you don't like them. You don't have to be there long to get a gut feel of the personality of a unit and the leadership style of the unit (Inspector, focus group 7).

> We'd [inspection team members] all have an opinion, but it would just be an opinion. I don't know how one would be able to objectively quantify an opinion about how a commander is leading in an [investigation] report (Inspector, focus group 4).

Inspectors also explained that assessing leadership within the context of any inspection would be problematic because it is a stressful and artificial situation in which leaders are likely not behaving in their typical fashion. As an inspector told us:

> We are only on site for three days, and only in a stressful situation, so I'm not sure how valid it is [to measure leadership in that context]. It's the day-to-day tools that would be better for it. People are on pins and needles during the inspection (Inspector, focus group 6).

A comment made by one of the leaders from a recently inspected wing reinforces this premise:

> In a week, I don't think you could figure it out. With enough smoke and mirrors, I could fool you for a week. But if you look at if the mission is being accomplished, you will see over time that a person is inept at getting the job done (Inspectee, leadership interview 9).

Further, inspectors raised objections to assessing leadership within the context of a compliance inspection. They felt that, especially compared to an ORI, a compliance inspection does not provide an opportunity to truly see what a leader does:

> For the UCI inspection, leadership is not a focus for us. . . . ORI is a different ball game. During an ORI, I do grade and look at leadership. In cyber inspections, leadership is not a focus (Inspector, focus group 8).

> The snapshot [of leadership] you get is better in operational readiness inspections. CIs are more difficult for gauging leadership; ORIs are better. There's no specific checklist, but it's a much more integral part of ORIs. Leadership is key in the success or failure [of the ORI] (Inspector, focus group 1).

We also heard in several focus groups that, during a compliance inspection, the emphasis is on the programs, not the people.

A final argument that inspectors offered to support their assertion that leadership should not be evaluated during inspections was that there are already adequate tools in place for this purpose. On more than one occasion, inspectors noted that the Climate Survey, discussed earlier in this chapter, provided honest, direct feedback on leadership effectiveness. However, as one inspector observed, it "is a commander's tool, the data are not for anyone outside the unit" (Inspector, focus group 6). Personnel evaluations were also cited as a source of data on leadership effectiveness.

Summary

The new UEI expects to assess wing discipline and leadership in ways that have not been attempted in the past. SAF/IG asked PAF to identify potential methods of measuring these two important wing attributes, which would receive a closer look than ever before under the new UEI.

Discipline is a multi-faceted concept. Air Force personnel in the field associate it variously with adherence to technical orders and following rules, customs and courtesy, wing member legal infractions, and even physical fitness. These differences of perception make it difficult to conclude which form of discipline should receive the most attention in an inspection. Air Force personnel also varied greatly in their view of how different aspects of discipline might relate to unit compliance or readiness. The scholarly literature has not given this topic much systematic attention, making it even more challenging to identify widely accepted or frequently used measures of discipline that could be adapted for UEI purposes. We found no definitions in other organizations that would be particularly useful to the Air Force. In net, our findings—or lack thereof—on measures of discipline were not supportive of their use.

In contrast, the information we found on measures of leadership was superior in terms of both quality and quantity. In our interviews and focus groups, Air Force personnel identified a number of leadership characteristics that have already been operationalized in AFMA's Climate Survey and AFSC's safety-focused AFCAST surveys. The Climate Survey collects information on personnel's perceptions of unit leadership, among other things, from all wings every two years. Units can use AFCAST surveys to assess their personnel's perceptions about the state of safety in their units, including the role and effectiveness of unit leadership in assuring safety. Methods developed by academics, such as the MLQ, psychological safety measures, and the use of data aggregation, provide additional and oftentimes complementary ways of assessing not only the effectiveness of a wing commander, but also that of the entire wing leadership chain of command.

Broadly speaking, these sources point to attributes of leadership that the scholarly literature tends to group into two categories: transactional and transformational. Transactional elements of leadership tend to focus on output goals, adherence to standards relevant to achieving these goals, and rewards for wing personnel that are clearly linked to the levels of output and standards a wing achieves. In some ways, it resembles definitions of military discipline offered by Air Force personnel and, as such, may be a useful way to operationalize both leadership *and* discipline. Transformational elements of leadership, on the other hand, tend to emphasize how a leader or chain of command motivates and even inspires wing personnel to work toward a common purpose by encouraging them to exceed their own expectations of what they can be and achieve together. Transactional leadership tends to be output-oriented, while transformational leadership is strongly people-oriented.

While both categories of leadership are important to compliance, their relative importance depends on the specific environment of a wing. Analytic literature and formal surveys offer many ways of measuring these types of leadership, and extensive analytic work has been done to validate a number of these systems of measurement.

The many available approaches to measuring leadership give the Air Force important options. The IG could simply decide to use information collected on a wing through the Air Force Climate Survey or AFCAST surveys as integral elements of an inspection that the IG could then combine with more traditional inspection material to assess a wing's compliance or readiness. Alternatively, if the IG cannot acquire satisfactory access to such survey data or they are not timely enough, the IG could use the structure and questions from these surveys to develop its own surveys for application in conjunction with major inspections. Or the IG could eschew such survey methods and use insights from them, the analytic literature, and Air Force personnel perspectives to develop checklist items inspectors could use to assess wing leadership in a more traditional way. Chapter Eight discusses such options further.

Although our analysis provides a strong case for measuring leadership within the context of a UEI, our fieldwork also suggests that there may be psychological barriers to overcome in the process. Based on their experience with the Air Force inspection system in place at the time of our research and, to some degree, its predecessors, many Air Force inspectors have reservations about assessing leadership during a compliance-focused inspection.

CHAPTER SIX

Introducing the Management Internal Control Toolset (MICT)

MICT is a "new computer-based inspection program . . . [being] billed as an all-in-one inspection tracking and analysis toolset. The program is web-based, real-time, and allows individual units and program managers to assess their programs and up-channel their internal inspection results" (Curry, 2009). The Air Force Reserve Command (AFRC) first made it available to its wings in October 2008, then began a concerted effort in January 2009 to introduce the program for use in all wings command-wide (Anderson, 2012).

At the time of our research, the Air Force planned to introduce the program to the active component beginning in December 2011 (Smith, 2011). The Chief of Staff Air Force Team Excellence Award program recognized and validated the AFSO21 initiative that led to the creation of MICT as an Air Force best practice (Abalo, 2009).

Implementation of MICT in the Air Force Reserve has moved at different rates in different units. It has been relatively smooth, even though some users have found it too complex. The "Directorate of Analyses, Lessons Learned and AFSO21, together with the Directorate of Financial Management, validated that MICT produces an annual 38.7 percent improvement over maintaining and operating . . . legacy systems or an estimated $12.1 million in valued efficiencies over the 880 units within AFRC. If implemented Air Force–wide, efficiencies could easily top $100 million annually" (Abalo, 2009).[1]

Several active wings examined MICT before the Air Force–wide roll out and found several elements useful. For example, the 305th Air Mobility Wing identified the following benefits (Anderson, 2011):

- It facilities oversight by a wing commander.
- It saves the time of those entering and using data in the system.
- Its checklists are easy to update.
- It is accessible and user friendly.
- It is flexible enough to accommodate all local inspections and exercises.
- It supports discrepancy trending, root cause analysis, and corrective action programs.
- It allows for the easy sharing of information managed in the system.

---

[1] The metrics emphasized in this assessment suggest that MICT would be valuable to the Air Force because it would (1) improve the quality and (2) decrease the cost of information management relevant to command and control of Air Force activities. Not surprisingly, the senior Air Force leadership and the personnel we talked to had other ideas as well. Different supporters named different potentially attractive attributes. In addition to the two metrics above, they spoke of the potential for improving wing-level decisionmaking, cross-wing benchmarking, and accountability of wing command staffs. Different priorities point to different ways of implementing MICT, a point that becomes more apparent in the following discussion.

Most people in the active component of the Air Force, however, still have never heard of MICT.

To learn more about MICT, we spoke with Air Force personnel directly responsible for implementing the program about the reserve component experience to date. We also used our focus groups and interviews to ask a cross-section of Air Force active component personnel what they thought about a new program *like* MICT. For the most part, the active component personnel gave us their perceptions of what they would expect from a new piece of software that would manage and report wing-level self-inspection information in a standard way across the Air Force.

In this chapter, we first offer broad observations on the potential pros and cons of adopting a standardized wing-level information technology system like MICT. We then report more specific concerns that the potential benefits of MICT may not extend to the wings that bear the costs. If this occurs, the quality of the data that MICT manages could be jeopardized. We also report concerns that, even if MICT can generate sufficient benefits at the wing level to offset its costs, users may not apply MICT in a way that realizes those benefits. Finally, we report concerns that sharing data from MICT with inspectors, MAJCOMs, or other Air Force overseers outside a wing could compromise the quality of the data the wing places in MICT.

Pros and Cons of a Standardized Wing-Level IT System

Potential Benefits

Our focus groups and interviews highlighted three principal potential benefits of MICT. First, inspectees believed that a single system used to manage all checklists and keep them up to date would provide a typical wing commander with a clearer picture of what compliance means in the Air Force. Such a system could, in effect, serve as a continuity book for a new commander. As one participant noted, "the Air Force doesn't educate commanders about what their responsibilities are." A system like MICT could help "commanders know what they were responsible for before we said, 'Here are the keys to the Ferrari.'" By clearly documenting what will be inspected at a wing, MICT helps a wing "create plans for continuity" during the transition from an old to a new commander (Inspector, officer focus group 2). When a commander first arrives at a wing, MICT would allow him or her to easily access information about all the things that are expected of the wing from a compliance perspective. MICT could also potentially highlight relevant trends in the wing and any areas that might require special attention.

Second, inspectors and inspectees believed that information managed in a standard format across all wings in a MAJCOM or across the entire Air Force could support cross-wing trend analysis. One respondent observed that analysts could "find similarities across wings or groups. If something is always compliant or non-compliant, you can see that and ask 'why?'" (Inspectee, leadership interview 5). Such analysis could identify the best performers and suggest where to look for best practices. It could also identify broad trends over time faster than any wing could by itself. Also, broader trends might warrant attention at a higher level than any one wing. A standardized system could, in principle, deliver relevant data to the appropriate level of the Air Force where responses could be made.

Third, inspectors and inspectees told us that it would be easier for a MAJCOM IG to audit a wing's self-inspection system if all wings managed data in a standardized way. The IG would know exactly what type of data to expect and where to look for it. One participant stated, "It would be nice to have a standard program so I know when I come to a wing where to find things" (Inspectee, leadership interview 8). Moreover, managing data digitally via a Web-based system creates the potential for a virtual audit of some or even most of a wing's self-inspection system.

Potential Challenges

Still, the majority of people we spoke to about MICT were skeptical. The greatest concern stemmed from a general skepticism about the promises that accompany any new software product. Often citing specific examples from their own experience, inspectors and inspectees questioned whether MICT's software would embody the benefits promised, whether the Air Force would invest enough in training and support to ensure that users could realize these benefits, and whether wing commanders would ultimately use the information generated by MICT to improve wing performance. We will address these points and others related to them in more depth in a moment.

Second, our participants wondered how easily MICT could accommodate local variation from wing to wing or across MAJCOMs. Inspectors especially asked if local users could customize it enough to capture their particular needs. Also, they wondered how much wings could customize MICT without defeating the purpose of a standardized system. Even if the software allowed simple customization, would Air Force culture allow it? Would the availability of a standard data management system encourage greater use of Air Force–wide standards, even if they degraded local performance? In sum, how would the Air Force balance the potential benefits of customization with the benefits of standardization?[2]

Third, inspectees wondered if MICT would be as effective as existing systems. Over time, many of these systems have evolved advanced capabilities. While IG inspections routinely seek local capabilities that can be identified as Air Force best practices and exported to the rest of the Air Force, recent local innovations often elevate some local capabilities above those available elsewhere in the Air Force. Therefore, it is highly likely that the wholesale replacement of such capabilities would degrade local performance in at least a few areas. However, if legacy systems persist side by side with MICT, then MICT cannot achieve its purported benefits of simplifying data entry and generally reducing the cost of managing a self-inspection program. As one leader summed it up, "If [the system] is truly a user-friendly and reliable system, you will get people to use it. The units are required to have some type of documentation, so they will develop their own program, their own Access database or Excel spreadsheets, if they can't use the one they are expected to use" (Inspectee, leadership interview 3). Any new standard system, like MICT, must be as good or better than the homegrown products that can be developed to replace them.

Finally, some personnel worried that the availability of a standardized system might encourage Air Force resource managers to reduce the number of billets available to external inspectors and assessors or to wing personnel tasked with inspection-related responsibilities (e.g., wing IG, QA personnel). They know that the Air Force is currently under tight resource

---

[2] As noted above, the nature of any preferred balance depends on the relative importance of competing goals and priorities. The Air Force has not clearly stated these goals or their relative importance and has no formal process in place that could resolve these issues quickly. In all likelihood, competing interests will shape MICT as implementation proceeds. Viewed from this perspective, the concerns in the text can be stated as uncertainties about what balance will ultimately emerge from this process.

constraints and that these constraints will likely get worse before they get better. It might be easy for managers to view MICT as a form of inspection automation that would free up manpower for use elsewhere, even if inspectors and assessors are still required to generate and use the data that MICT manages.

Will MICT Benefit Local Users More Than It Costs Them?

Suppose we give MICT the benefit of the doubt on the points above and assume that it has the potential to generate more benefits than costs. There is still the question of who will benefit and who will pay. For example, suppose the principal benefit is the creation of new cross-wing trend analysis, but this benefit is available only if personnel within wings assure the quality of data in MICT, and quality assurance distracts them from other priorities within the wing. Can MICT really create an Air Force–wide benefit if the wings responsible for managing it do not see much benefit for themselves? For example, one interviewee said, "We already spend lots of time creating spreadsheets that aren't helpful to us but are helpful at the higher level. It sounds to me like more work that I, as a captain, won't ever use" (Inspectee, leadership interview 2). This is a classic problem with the assurance of quality in any data system. Our fieldwork participants rarely stated this concern in such direct terms, but they offered us many examples of this kind of problem. Taken together, their responses question whether MICT will generate enough local benefits to offset the costs associated with its local management. If it does not do this, MICT will probably not realize its potential. Evidence from our discussions, interviews, and focus groups with Air Force personnel tells us that MICT is more likely to realize its potential the closer it comes to meeting certain criteria.[3]

First, MICT must be mature. For one airman, this meant that,

> [I]t should be effective right out of the gate. Uncle Sam will put out software that's not tested. Then it crashes and he [Uncle Sam] spends more money fixing it than he initially paid for it (Inspectee, focus group 5).

In broader IT practice, this means that MICT should have few bugs and those that remain should rarely affect standard operations. The current version of MICT reliably performs the data management tasks it was designed to execute without requiring workarounds. Patches used to overcome past problems are not apparent; the version of MICT in place today operates as though those problems never existed. When a user makes an error, MICT will be forgiving. A failure to enter one piece of data properly will not endanger data entered properly beforehand. When a user makes an error, MICT will recognize the error and alert the user.

Air Force personnel also indicated that MICT must be simple to both learn and use. The current version of MICT is designed from the user's point of view so that user actions are intuitively appealing. One leader told us, "I am not computer-savvy, but I can do TurboTax myself, because it's easy to use, intuitive" (Inspectee, leadership interview 3). When a user is confused, MICT offers quick access to clear instruction. Ideally, that instruction includes access to a real human at a help desk who can walk a user through any standard task associated with MICT.

---

[3]   These criteria do not reflect RAND's analytic judgment. We assembled them from specific statements of many Air Force individuals about what they would want from a new standardized wing-level information management system.

Documentation is clearly written—again, from a user's point of view. Training is also offered from a user's point of view and has been matured to reflect learning from earlier users. This training is available to any person who will use MICT and imposes reasonable burdens in terms of time, financial cost, and intellectual challenge. The training and documentation are designed with real users in mind. The Air Force provides adequate access to training and on-the-spot assistance as users continue to turn over in normal rotation patterns. One leader told us,

> I spend more time doing IT training than visiting my airmen. I spend two to three hours every day doing IT stuff. Keeping tools up-to-date costs manpower. It is more things out of hide, and we are already facing manpower cuts (Inspectee, leadership interview 11).

Fieldwork participants felt MICT must be introduced with a robust, well-planned transition strategy. Legacy systems typically remain in place until users are confident of how the new system works. Parallel use of legacy and new systems is facilitated by systems that allow simple point data entry and manipulation, but if a user cannot understand the new system at any point, the legacy system remains available as an option. If the new system fails during the transition—because of flaws in the system itself or user errors—the transition plan mitigates the effects of this failure, perhaps by using the legacy system(s) as backup. Once users are confident with the new system, the legacy systems can be shut down without any complex closeout operations to secure or transfer data.

Once in place, MICT must be robust in the face of reasonable operational changes. The current system continues to effectively support a wing even if elements of the wing deploy or are detached, or if new operational capabilities are added to the wing. MICT seamlessly accommodates the movement of equipment and personnel during routine deployment cycles. The system can also mitigate the effects of a wing's loss of knowledgeable personnel because it is easy to use and easy to learn. Also, if a wing sustains enough familiarity with MICT, knowledgeable users will abound. However, one strong skeptic told us,

> You'll drop this on me without teaching me. You'll give me another database and I'll have to find "the guy" that can manage it, and who will never deploy; he'll just manage this. Just like Windows 7. No one came to explain and now we can't print. And like the Defense Travel System. Either it will work perfectly, or I'll be dead in the water. The problem is that we don't train people to use the technology, and we don't train them how to get along if it goes wrong. There is software that tracks whether pilots are currently qualified. But if the computer crashes, no one can fly (Inspectee, leadership interview 6).

As this comment suggests, when MICT fails, which participants perceived as inevitable, it will meet better long-term success if users are adequately trained in a back-up system so that the operational capability of the wing is not at risk until the system comes up again.

MICT must not simply "add another layer" of data management to a self-inspection program. As one NCO noted, "We already have tracking tools. Honestly. We don't need another program to try and replace or cobble together with the different programs that we do have" (Inspectee, NCO focus group 2). This will mean that MICT should either (1) completely displace legacy data management systems—at least following full transition to the new system—or (2) be fully and seamlessly integrated with existing legacy systems so that relevant actions need only be taken once.

Our interviews with personnel responsible for developing MICT and interaction with ISITT members indicated that MICT must also be able to accommodate any local self-inspection requirement. The Air Force as a whole and each MAJCOM can identify data they need to support their own trend analyses or other oversight from above the wing. MICT will support the data requirements of decisionmakers at these levels, but it can also support local operation, including unique capabilities, unique priorities, and local innovations that do not warrant diffusion to other wings. In particular, it will allow a wing to track compliance with elements of an Air Force instruction not highlighted on an Air Force– or MAJCOM-level checklist.

Moreover, MICT should support full root cause analysis. Even if an inspector rarely pursues every symptom to a root cause that a wing can fix, a self-inspection system exists precisely to identify and eliminate problems proactively—before they affect mission, safety, or other compliance priorities of the wing. Merely managing data is not enough. MICT must facilitate the application of that data to problem solving. This presents an inherent question about how best to enter data. Fully automated data entry, using carefully fleshed out menus, limits the number of keystrokes required to enter data and facilitates the analytic manipulation of the data entered. Full text entry, on the other hand, allows for a more subtle documentation of problems that may support root cause analysis, especially when the problems are new and so not reflected in existing menus. Root cause analysis will be most effective if MICT is structured to strike a balance between (1) menu-driven speed and simplicity and (2) the subtlety and flexibility that free text allows. This will likely require ongoing contractor support to facilitate learning and refinement of the system.

As we understand it, MICT can potentially address each of these concerns.[4] While it is still being developed, its basic structure is relatively mature. It is also relatively easy to learn how to use, especially for younger personnel who are more conversant with technology products. Training requires only limited effort. MICT can support the automated transfer of data, which facilitates a smooth transition. It appears feasible to fully replace most legacy systems without losing functionality. MICT is flexible enough to accommodate broad variation relatively easily and it can accept data entry in multiple forms. The responses we heard in the field tell us that the Air Force will achieve a better reception for MICT if its leadership first addresses the wings' concerns head on. At least in principle, the leadership can take advantage of MICT's basic robustness and flexibility in designing an implementation program that addresses these concerns.

Will Wing Personnel Use MICT Appropriately?

MICT is first and foremost an enabler. Human input is necessary for it to reach its potential. We just discussed those things that Air Force decisionmakers can do to most effectively implement MICT. This section focuses on what users of MICT can do to maximize the system's potential.[5] This is where *sociotechnical concerns*—issues about the interaction between infor-

---

[4]   Among the many contributions made by one of our reviewers, Michael Greenberg, was this succinct and apt summary of these concerns: MICT is more likely to fulfill its potential if it is: (1) bug-free, (2) user-friendly, (3) robust, (4) non-redundant, and (5) responsive to wing-level operational demands (e.g., deployments).

[5]   The statements offered here do not reflect RAND's analytic judgment. We assembled them from specific statements of many Air Force individuals about what they would want from a new standardized wing-level information management system like MICT.

mation technologies and the people who use them—really come to bear.[6] Based on their past experience with related issues, our respondents voiced considerable skepticism about whether future users will use MICT properly. This ultimately helps explain some of their broader skepticism about the benefits that MICT's proponents currently promise.

The wing commander holds the key to the successful application of MICT at a wing. The chain of command at a wing is sensitive to the priorities of the commander. If the commander takes the data in MICT seriously, other leaders in the wing will as well, and airmen will follow suit. A commander can show interest by making reports from MICT a regular item on his schedule. He or she can play an active and visible role in assessing corrective action plans for deficiencies identified in MICT and deciding when corrective plans are complete and ready to be closed out. He or she can also reward members of the chain of command for proactively identifying problems, especially if they are held accountable for resolving those problems quickly. That is, he or she can shift the focus of rewards and sanctions from the *revelation* of problems to their *resolution*. Some of the Air Force personnel we spoke with took this even further. They suggested that a wing commander can tie MICT to wing *esprit de corps* by identifying examples in which MICT has provided information needed to improve the wing's performance, particularly if that improvement can be linked to greater mission capability. This helps close the loop in the oversight system enabled by MICT by demonstrating that bottom-up efforts to expose and resolve problems lead fairly directly to positive outcomes for the wing as a whole. The wing commander can maximize this benefit by mobilizing each link in the command chain to use MICT to achieve real improvements at the wing.

People with access to information have a temptation to act on it. Pushing information through an organization from the bottom up can invite micromanagement. Our respondents emphasized that MICT should not be allowed to displace the chain of command. Decision-makers at lower levels will always continue to have information that those higher in the chain of command do not have. If it is accurate, the information in MICT may appropriately push some decisions further up the chain of command, but it can also give senior leaders in the wing the information they need to hold subordinates accountable for solving the problems that exist at their levels. Because MICT gives the commander access to more information, s/he can, in principle, devolve *more* responsibility without taking more risk. Such delegation of responsibility through a well-functioning chain of command presumably helps build the capabilities of more junior commanders, a key task of the chain of command itself. If senior leaders misuse the information that MICT gives them, however, they can actually degrade their development of the young leaders who one day will replace them.

Air Force personnel we met with also worried that senior leaders might react to improved information from MICT in another negative way: they might conclude that the face-to-face contact they used to sustain effective oversight of their subordinates in the past is no longer necessary. The easier it is to absorb high-quality information from MICT, the more tempting it may be to lean more heavily on that information than on the more qualitative information that flows through face-to-face contact. This is especially true when demands on a leader's time increase and s/he looks for places to economize his or her effort. These respondents placed special emphasis on MICT's role as an enabler and the importance of direct contact to effectively control a chain of command. According to one person, MICT cannot be "a replacement for

---

[6] For a useful overview of this perspective, see Bikson and Eveland, 1998.

human interaction. Any commander that doesn't get out and learn what his people do in any instance is shooting himself in the foot. But if he has a tool right in front of him to measure your progress . . . he shouldn't have to come out and check your checklist." In response to this comment, another noted, "*If* it was controlled that way, fantastic."

Many participants asked why, even if MICT completely displaces the data management systems in place today, we should expect users to employ MICT differently than they have employed these legacy systems? In the past, they have not regularly updated the checklists used in legacy systems to reflect changes in policy at the Air Force and MAJCOM levels, they have not aligned their checklists to the checklists that their contract suppliers of services use in their own operations, they have not verified that the checklists relevant to every part of the wing in the legacy systems are kept in sync with current requirements and activities throughout the wing, and they have not regularly audited the data in their legacy systems to verify that they are accurate and up to date For MICT to achieve its full potential, leaders must address the lack of attention that (apparently) many give to the quality of checklists and data in the data management systems being used today.

Sharing Data from MICT Outside the Wing

Official Air Force statements released just as AFRC began rolling out MICT across the command echo the following view: "Once self-inspection results are entered and submitted, local wing, numbered Air Force and AFRC inspection monitors may see the results in a nearly virtual real time state. Commanders from the highest level on down will be able to rapidly assess inspection progress results and areas for improvement" (Curry, 2009). Creating such broad visibility into the inner workings of wings has been an integral part of the appeal of that MICT has for its advocates. Yet, our respondents almost uniformly rejected this use of MICT. They prefer that it remain primarily a tool to be used by each individual wing commander and believe it will be most effective if the data it records remain primarily within the wings from which they are collected.

The interview and focus group participants were somewhat more willing to share data with outside IG personnel than with others above the command. Giving the IG access to MICT data would presumably simplify inspections and, by making more IG oversight virtual, reduce at least the logistical burden of outside inspections on wings. By giving inspectors advance information on points of weakness in a wing, MICT data could help inspectors focus on the wing activities most likely to benefit from their attention. However, even if this proved to be true, various respondents raised two concerns.

First, as explained in Chapter Three, most inspectors prefer to come to a wing with an open mind. Access to MICT would necessarily predispose them to finding certain circumstances. Inspectors worried about this effect might even decide not to examine MICT data before they arrived, just as they avoid looking at wing-specific compliance data before an inspection today. If this occurred, MICT data would have a limited effect on inspection preparation and hence on the burden of external inspections.

Second, airmen who know their data will be aired outside a wing might be reluctant to submit accurate data. Senior leaders within a wing might even be reluctant to encourage them to do so. Requiring the sharing of MICT data outside a wing could compromise the quality of the data in MICT and thereby reduce their usefulness to the wing commander seeking to make improvements within the wing. In the words of one respondent, "Clearly, what you say internally to your unit is different than what you say externally. . . . You need to not be delivering bad news" (Inspectee, leadership interview 8). It is well known that attaching rewards or sanctions to self-reported data can, and often does, compromise the integrity of those data.[7] The FAA has learned over time to help ensure that data are accurate by using collection methods that protect the people reporting them from consequences. In effect, our participants suggest that negative consequences are less likely if MICT data remain within a wing. This reflects an implicit expectation among our respondents that external inspections will not reward accurate reporting and then hold wings accountable for correcting problems revealed expeditiously. Instead, they expect external inspections will continue to sanction wings that allow the problems to occur in the first place, even if they self-report them.

Our respondents felt more strongly about sharing MICT data with MAJCOM and Headquarters Air Force functional and command staffs. Because reporting problems could bring negative consequences from above, wing personnel would be less likely to reveal problems in their MICT data. Without access to such information, wing leaders could not identify and resolve problems proactively. In addition, some respondents feared that, just as greater visibility within a wing might discourage the wing's senior leaders from devolving responsibility for appropriate decisions, releasing data outside the wing could encourage leaders above a wing to micromanage affairs within the wing. As one respondent put it, a data report in MICT will provoke a query from above. When the wing responds to that query, a tasking will follow. "If they send me a tasker every time I put something up there, I'll stop reporting. I'll keep track, but I won't press 'send'" (Inspector, officer focus group 1). This suggests that both the initial query and the follow-on tasking will place a direct administrative burden on the wing and transfer authority that is best retained within a wing to a higher level with less relevant information. Another person worried that if a directive from above the wing "does not have sufficient command influence, then the units can spin themselves [by] doing something not important to the wing. The wing-level priorities are based on wing tempo" (Inspectee, leadership interview 9). He worried that functional priorities could easily drive directives from above the wing. To avoid the costs associated with such events, personnel within a wing might avoid reporting negative findings in MICT—again compromising MICT's usefulness to the wing commander.

Several Air Force personnel anticipated these problems but also appreciated the potential value of sharing MICT data outside a wing, especially with the inspector general. As a compromise, they suggested that MICT data should normally remain within a wing, but that external inspectors might be granted short windows of access to help them prepare for inspections. MICT has a fairly sophisticated ability to assign specific permissions for viewing each piece of information it contains. By applying those permissions appropriately, a wing might be able to limit the information external inspectors can see to that which covers a specific period of time. But once access to any data is granted, those data are permanently available to outsiders. Moreover, any effort to restrict access limits the usefulness of MICT data to the Air Force

---

[7]   For a discussion of this point, see Stecher et al., 2010. Organizations seeking to avoid this problem often carefully separate data that will be used to incentivize personnel ("motivational metrics") from data that the leaders will use to assess the status of the organization ("diagnostic metrics"). Or they develop measurement systems that their personnel cannot game or shape to their advantage. Doing this is typically quite challenging.

as a whole. Limited access to select windows could have an especially deleterious effect on the ability to track trends over time at the MAJCOM level or above.

Summary

When interpreting the perceptions reported in this chapter, it is important to keep in mind that few of the Air Force personnel we talked to knew much about how MICT actually works in practice. Their comments typically reflect their own past experience with efforts to use new information technology tools to improve processes. Through these experiences, they have learned not to expect everything promised by such innovations. More specific knowledge about MICT would probably allay some of their concerns. Experience in AFRC suggests that MICT shows great potential for easing the management and use of self-inspection data. The design of MICT software, however, will likely be only one source of concern among many. While many accept that MICT could improve their lives, others question whether the Air Force will, in fact, implement and support MICT in a way that allows the active component to benefit from its full potential. Will users apply MICT in ways that can effectively inform decisionmaking within and above wings? In particular, will users be willing to populate MICT with accurate data, even if some of those data point to problems that require substantial corrective actions to fix? These questions are directed less toward MICT per se than toward how the Air Force will manage MICT after it is introduced. Growing resource constraints raise particular concerns about whether the Air Force will resource MICT appropriately or will instead see it as a way of automating leadership and inspection tasks that are better conducted in person.

CHAPTER SEVEN

Implementation of Significant Change in the Inspection System

The previous chapters provide information about five different elements of change in the Air Force inspection system currently under discussion. Implementation of some of these changes has already begun. Other changes need to be defined more clearly before formal implementation can begin. Even if the Air Force ultimately decides to accept all these changes, they cannot occur overnight. Initial experiences with the new systems will lead to potentially significant adjustments. Successful basic implementation across the entire Air Force is likely to take a long time.

This chapter summarizes what has been learned from efforts to achieve similarly significant changes in large, complex American public and private sector organizations over the past three decades. It then offers a brief history of how changes like those under discussion in the Air Force recently occurred in the FAA's inspection system for commercial aircraft.

Recent Formal Change Management Perspectives in the United States

The quality management movement that came to the United States from Japan in the 1980s and has grown and prospered ever since brought with it an approach to formal change management that large, complex public and private sector organizations have used to successfully implement a wide range of changes.[1] The approach has been studied and documented in publications aimed at academic, technical business, and broader popular audiences.[2] John Kotter became especially well known in the 1990s for the conclusions he drew from an extensive practice helping large corporations implement change (1996). This section draws on Kotter and others writing in the tradition of the quality management movement to summarize the important elements of formal change management. There are few substantive differences among these writers. Rather, they tend to emphasize different aspects of formal change management. Readers interested in a particular element of the change process should consult the authors who give that element greatest emphasis.

We will break down the change process into three elements—planning, execution, and sustainment. Some authors refer to these as steps or phases. Continuous improvement is particularly important to the quality movement. In such a setting, it is best to think of change as

---

[1]   For more on the link between the quality management movement and current American thinking about formal change management, see Camm, 2003.

[2]   For literature reviews of academic and technical business publications likely to be useful in an Air Force setting, see Zellman et al., 1993, and Cook, Castaneda, and Haddad, 2010.

a series of increments, each building on the last by monitoring the previous increment's performance and adjusting the next to benefit from what is learned from the empirical performance data. At any point in time, a large organization will be planning some incremental changes, executing others, and sustaining or institutionalizing still others. Information will be flowing between all these efforts all the time. So, even though we will describe these pieces as though they occur sequentially, in fact, the pieces often have a great deal of overlap.

This dynamic view of change seems well suited to the Air Force context. The senior leadership at the Air Force, MAJCOM, and wing levels turn over on a fairly regular schedule. Even so, the Air Force can look out decades into the future to implement new programs and weapon systems. The Air Force is used to structuring these large changes so that leaders can easily pick up an effort already in progress and hand it off to the next team at the end of their tours.

The management of an aircraft program offers a simple illustration of such a change process. At any point in time, a complex aircraft program has multiple blocks at various points in their life cycles. Some blocks are assembling technologies that will make future capabilities feasible. Other blocks are in development and testing. Those blocks approaching operational release are in production. And still others are in the fleet, where the Air Force monitors their performance and adjusts support plans as they age. This is very similar to how the formal change management approach works. In fact, advocates of that approach could learn a great deal from how the Air Force (and the rest of DoD) has managed weapon system programs for decades.[3]

Table 7.1 summarizes the activities of the formal change management process that we will discuss. We have grouped these activities into the categories of planning, execution, and sustainment. The columns indicate representative references from the literature. We show Kotter first to reflect his dominance in this literature. We list the others in order of publication date. Camm et al. (2001), Moore et al. (2002), and Cook, Castaneda, and Haddad (2010) explain how to apply the formal change management approach in three very different defense settings. Fernandez and Rainey (2006) present the concerns of formal change management in public-sector organizations in general. Judson (1991) and Galpin (1996) are more general practical manuals on formal change management that complement the other publications shown here. The table shows which activities each of these publications emphasizes.

Element 1: Plan

Planning raises the "chicken-and-egg" problem that is present in much discussion of organizational change. Can significant change begin before gaining the support of senior leadership? If not, how does any idea for change ever come to the senior leadership's attention? Our analysis suggests that change is the product of an iterative process in which small teams design changes, test them on a small scale, bring the empirical results of these tests to some leadership group's attention, and garner support for further design and testing efforts that are ultimately the foundation for a significant change agenda backed by the leadership. This iterative process involves considerable cycling through the activities outlined below until enough support exists to move a significant change from the planning to the execution phase.

---

[3]   The Air Force Planning, Programming, Budgeting, and Execution System has a similar and completely analogous structure. The budget generated by each cycle is a discrete, incremental product of the longer-term process. At any one point in time, elements of planning, programming, budgeting, and execution activities in different cycles proceed simultaneously, constantly informing one another.

Table 7.1
Elements of Formal Change Management Emphasized in Recent Publications

Activity		Kotter (1996)	Judson (1991)	Galpin (1996)	Camm et al. (2001)	Moore et al. (2002)	Fernandez and Rainey (2006)	Cook et al. (2010)
Plan								
	Assess the current "as is" situation		X	X				
	Establish a compelling need for change	X		X	X	X	X	
	Create a coalition of relevant stakeholders	X		X	X	X	X	X
	Design a vision and strategy	X		X		X		X
	Create a specific plan			X	X	X	X	X
	Test and measure			X	X	X		
	Communicate	X	X	X		X		
Execute								
	Issue new policy							X
	Resource					X	X	
	Train				X	X		X
	Measure				X	X		
	Motivate	X			X	X		
	Build on incremental successes	X	X		X		X	X
	Communicate				X	X		X
	Consolidate successes	X	X				X	X
Sustain								
	Resource					X		X
	Train				X	X		X
	Measure			X	X	X		X
	Motivate			X	X	X		X
	Adjust			X	X	X		X
	Communicate				X	X		
	Consolidate and anchor	X	X				X	

Assess the current, "as is" situation. The standard quality management approach to change documents the current "as is" state of an organization; posits a future "to be" state based on the goals of change; and performs a gap analysis of the difference and what is required to get from the "as is" to the "to be" state. Documenting the "as is" state is the natural first step in such analysis. Experience has shown that this activity is necessary. Without it, we cannot know what the current state of an organization is or, perhaps more important, why it is in that state. Without knowing "why," we cannot effectively anticipate where resistance to change will occur and how to defang it. That said, a common lesson of change management is that large organizations typically spend too much time on this first step. An organization's "as is" state is something we can come to know with confidence during the change management process, even if we don't know it at the beginning of that process. Spending more and more time on getting to know an organization's current state provides a, perhaps unconscious, excuse not to start the much harder process of mapping what change might actually look like.

Establish a compelling need for change. Change managers have learned through hard experience that significant change typically does not succeed if it is harder to move forward than to go back. Because organizations are typically deliberately structured to preserve the status quo—that is why they exist—serious change raises basic existential issues about processes and procedures that current employees have often created and committed themselves to (again, often unconsciously) emotionally. Successful change requires a compelling explanation for why it is easier to move forward. To be effective, that explanation should be grounded in reason while also having an emotional impact. A metaphor often used is one of a "burning platform." Change managers seek to convince those individuals who must change their behavior in order for organizational change to succeed that they are on a burning platform and will go down with it if they do not find another place to stand.

Create a coalition of relevant stakeholders. Individual people change organizations. Individuals must change their behavior in specific ways for organizational change to succeed. An early task of any major change effort is to determine which individuals must change their behavior for organizational change to succeed and who within the organization represents their interests or has authority to direct their behavior. These stakeholders must sign on to the change if it is to overcome resistance. Again, organizations are mainly designed to preserve the status quo, and each stakeholder has a current role to play in doing that. The primary stakeholders have been around for a long time and know how to exploit the current system to their own advantage. Change often fails when those stakeholders who oppose change hunker down and wait for the advocates of change to give up their mission.

Large change efforts benefit from formalizing coalitions at two levels. The executive level (in the Air Force, the general officer and senior executive service [SES] level) provides top cover for the change effort and monitors it regularly to ensure that it continues to advance the broader goals of the organization. If the change is not achieving the gains promised, this executive group has the authority to change its direction or shut it down. The management level (in the Air Force, the O-6/GS-15 level) works the details. It ensures effective day-to-day coordination among all MAJCOM command and functional communities within the coalition. It designs the details of the change, oversees its execution, and reports on progress to the executive level. Coalitions at both levels must remain intact and functional for the duration of any successful major change.

Design a vision and strategy. The vision and strategy summarize the need for change, the shape of the change, and the roadmap to achieve it in the simplest terms feasible. They should take the form of an "elevator speech" that identifies the organizational goals in play, a short list of metrics for success explicitly linked to these organizational goals, and who will be held accountable to achieve these goals. They should be designed so that these basic elements will endure over the whole life of the change. Their content is the currency of the executive coalition overseeing the change. They should appeal to both the reason and the emotions of the stakeholders who will oversee the execution of the change.

Create a specific plan. This is a detailed campaign plan that implements the vision and strategy. Its content is the currency of the management coalition overseeing the change. It assigns specific roles and responsibilities, sets schedules, identifies and distributes resources, and details all the actions described below that will support change as it proceeds. It is designed to be malleable. Within the strictures of the vision and strategy, it regularly adjusts in response to accumulating experience with the change. No campaign plan survives the first contact with the enemy; the same is true here.

Test and measure. Empirical measurement and assessment is an integral part of the broader quality management approach. Pilot projects offer early opportunities to do this in large change efforts. They allow change managers to test hypotheses about alternative designs. They present early evidence on the validity of a change effort that managers can use to sustain higher-level support. They identify weaknesses in a setting where the negative effects are relatively easy to mitigate and learn from. Politically, pilots also offer one way to help secure ongoing support within a coalition, especially if the members of the coalition can propose pilots that address their concerns.

Such pilots should occur early—as soon as enough support exists to define a design worth testing. They should avoid the temptation simply to *demonstrate* a proposed change without giving it a rigorous test. But they should occur quickly, trading some rigor for the value of generating feedback early enough to inform ongoing decisions. Pilots can be viewed as small, incremental changes in their own right—at least for their duration. Effective pilots follow guidelines very much like those described here for larger and longer-term incremental changes. The degree of oversight and support of the pilots is scaled to their size.

Communicate. Formal change management in large, complex organizations has many moving parts, each one serving somewhat different priorities and goals. Constant communication among all these parts—from top to bottom, bottom to top, and side to side, and in multiple media—ensures that participants share as common a picture of the change effort as possible as it progresses forward. By its very nature, significant change ensures that information becomes obsolete as change proceeds. Effective communication mitigates this degradation, making it easier to keep all the moving parts aligned to a common, synchronized purpose.

Element 2: Execute

Execution in effect elevates an incremental change from the status of an idea to that of a concrete change in an operational setting. The oversight efforts and resources required to facilitate such an operational change are greater than those involved in the planning stage. Although all the issues discussed here are also relevant to planning, their scale as it pertains to execution becomes large enough to give them more explicit and detailed attention.

Issue new policy. Those who must change their behavior for an organizational change to succeed must know what that change is and what it means to them. For example, how does the change affect their roles and responsibilities? How does it affect their status within the organization? How is their behavior supposed to change? How will their change in behavior benefit the organization? What consequences will they face personally if they resist change?

Resource. One of the most common sources of failure in change efforts is a failure to support change efforts appropriately. Such efforts fail to appreciate the commonplace expression that "you have to invest money to make money." Successful change efforts tend to program explicit resources to support the change. Individuals changing their behavior are not expected to use their own funds to support organizational change unless they can expect to get offsetting funds if it succeeds. They are not expected to take on the change initiative without reducing other workloads unless they can anticipate future rewards from success that make it worthwhile in invest their own time in its success. In particular, they get relief from their normal workload to take time to participate in relevant training programs.

Train. Three kinds of training are important. One explains to individuals involved in the change why the change is good for their organization and, by implication, for them. The second explains the concrete details of the change, their role in it, and any new substantive skills they will need to execute their new role successfully. The third provides broader instruction on how to succeed in an environment where continuous improvement is the norm. This third form of training helps employees understand how individual incremental changes fit together and build on one another. It trains them to participate more effectively in the ongoing change process by observing the effects of change in their own localities within the organization and sharing this information to support ongoing efforts to build future increments of change. Training typically focuses within an organization, but organizations that perceive themselves as active links in integrated supply chains may also train individuals in the organizations they deliver outputs to and in the organizations they receive input from.

Measure. Successful change identifies metrics and tracks performance against these metrics. With the belief that what gets measured gets done, successful change managers seek metrics that align the decisions and behavior of individuals in the organization with the organization's higher-level goals. They also choose metrics that can inform them about how well the organization as a whole is performing, even if it is impossible to link such metrics to specific individuals in the organization. The campaign plan for a change sets targets for these goals on a schedule and tracks the performance of the change effort itself against such targets. Feedback from this tracking can inform the executives in the coalition overseeing the change. If the metrics are not helpful to them, they can ask for changes in metrics or the dates at which they expect target goals to be achieved.

Motivate. Effective motivation includes two steps. The first simply gives individuals room to change their behavior to promote organizational change. If the organizational change calls for them to sample risk assessment information, for example, it includes provisions that make sure that such risk assessment information is available. If it asks them to interview airmen to assess the quality of discipline at a wing, it gives them tools they can use to structure those interviews and the responses they get. Some writers refer to this first step as "empowerment" of those who must change their behavior.

The second is to motivating personnel to promote the organization's goals. Every organization has its own approach. Some use moral suasion, professional standards, and peer pressure. Aspects of transformational leadership fall into this set of tactics as well. Others adapt a style more consistent with transactional leadership, using tangible rewards tied to measured performance, which can range from direct cash payments, to larger operational budgets, to promises of promotion or better assignments, or to opportunities for career-enhancing training. The Air Force uses variations of most of these rather than options that directly involve cash payments and budgets. The Air Force could align the success or failure of any incremental change to the constellation of instruments that it already uses to motivate its personnel to promote its other goals. Successful change leads to good outcomes for those held accountable; failure to change leads to bad outcomes.[4]

Build on incremental successes. Significant change takes time. Large organizations are typically more impatient than the formal hurdle rates they are supposed to use to make decisions would indicate. As a practical matter, in private industry, leaders are often judged against quarterly goals or other relatively short-term metrics. In the Air Force, leaders turn over often, making it difficult for them to take credit for anything that takes longer than a year or so to achieve. Bite-size increments can scale a large change down into pieces small enough so that information about progress can be generated every few months. Successful change efforts track their performance at short intervals and trumpet news of success to those they rely on for continuing leadership attention and resource support. Such efforts start with modest resources on relatively easy tasks and, as a change team builds experience, confidence, and success, take on increasingly challenging tasks made possible by increasingly large commitments of resources.

Communicate. What was important during preparation for change is even more important during execution, because the resources—and so the stakes—are higher. Also many more, increasingly diverse players are involved. Tasks become more and more challenging. The need for active coordination grows. Rich communication, from top to bottom, from bottom to top, from side to side, and through many media, is critical to effective coordination as change effectively makes everyone's understanding of how the Air Force works out of date.

Consolidate successes. As sets of increments meet enough success for the organization to rely on them, legacy processes that are no longer needed can be dissolved. When all legacy processes are gone, the change is essentially complete. Formal change managers have differing views on how to manage this transition from the old to the new. Some advocate maintaining old and new side by side (known as "scaffolding") until the organization is absolutely sure the new systems will work. They promote this even if it creates duplication, confusion about which is the official system at any point in time, or ambiguity about the depth of the leadership's commitment to real change. Others see the rapid dissolution of legacy processes as a way to heighten individuals' sense of being on a burning platform or to convey the leadership's unbridled commitment to the future.

Element 3: Sustain

Two kinds of sustainment activities are important. The first concerns the institutionalization of the changes completed through any set of execution activities like those described above. Institutionalization transforms the completed changes from innovation activities into routine activities, i.e., a new status quo. Below, we will call the activities that result from such changes "now routine products of a formal change process." Institutionalization is likely to involve transferring responsibility for these completed changes from the coalition of executives and

---

[4]   The quality management movement tends to favor rewarding teams over rewarding individuals. But this presumes that work is organized to be executed by teams, sometimes without formal leaders, and that the teams have formal authority to act without concurrence of the organizations that its members come from. This makes it feasible to hold the members of teams accountable for their measured mutual performance. That is not how authority is assigned or how performance measurement affect performance assessment in the Air Force today.

managers coordinating the changes to the standard MAJCOM command or functional channels that coordinate most activities in the Air Force. The second kind of sustainment activity keeps in mind that, even following institutionalization, new increments of change are likely to continue, building on those that have been completed. The successful design of the new increments is likely to depend on a continuing information flow between them and the changes that came before.

Resource. Resourcing now occurs through standard channels and must compete for support without as much special leadership focus. This can place significant demands on organizations responsible for managing the now routine products of a formal change process. These activities continue to generate information that flows into ongoing change management efforts. These activities may retain personnel with expertise in successful change management who can be called upon as subject matter experts or trainers of the personnel still participating in ongoing change efforts. Ongoing change management efforts are likely to have better access to inputs from now routine products of a formal change process if the ongoing efforts program for their participation. Alternatively, the Air Force can encourage personnel involved in such now routine activities to make their information available to ongoing change efforts.

Measure. Organizational changes that result from quality change management efforts often build in a different kind of performance measurement than that used elsewhere in the Air Force. The measurement emphasizes ongoing benchmarking against analogous activities elsewhere and continuous improvement rather than the simple achievement of long-standing Air Force standards. Information from such now routine activities will be more valuable to ongoing change efforts the more it flows from this new approach to measurement. Such changes in measurement culture can also potentially seed broader application of this approach across the Air Force.

Motivate. Now routine products of a formal change process operate within the standard systems the Air Force uses to motivate performance. If the Air Force wants these products of change to continue to inform ongoing change efforts, it must ensure that it gives personnel in these activities space to participate in ongoing change activities and rewards them appropriately (in the context of their current primary duties) for effective participation.

Adjust. Effective, continuous monitoring of any activity is likely to reveal ongoing opportunities for improvement. The desirability of organizational change need not end simply because the Air Force now uses its standard oversight mechanisms to align the activity to the Air Force as a whole. As new opportunities for improvement come to light, activities discussed above with regard to planning and executing change return. If the additional proposed changes are large enough, it may be worthwhile to pursue them under the umbrella of the formal change management coalition described above. The effort to manage additional change should be scaled to the size of the additional change considered.

Communicate. As long as change continues, effective communication can keep all the relevant players up to date on the current status of the now routine products of a formal change process. In general, the quality change approach encourages greater communication across an organization than the Air Force as a whole does today. Preserving the tradition of such communication concerning now routine products of a formal change process could help seed this approach across the Air Force.

Consolidate and anchor. Over time, new participants in an activity that has used a new approach successfully for a long time come to the activity with no knowledge of what the activity looked like before the formal change management exercise. The new approach is fully internalized and taken for granted. People forget that significant change ever happened. As some have said about the gradual absorption of quality management methods into day-to-day management activities in successful organizations, people forget that the language they speak every day is called "prose." As noted above, that does not mean that change ends within the activity. It means only that, when current personnel involved in the activity today happen across the language used for the activity prior to the change management exercise, it sounds archaic and nonsensical to them. At this point, change is truly complete.

A Case Study: The Evolution of Policy and Implementation of Change in the Federal Aviation Administration Inspection System

The current FAA system of safety systems is the product of over five decades of thinking, legislation, regulation, and coordination among the government, private companies, and employee unions. The system has evolved as these three groups found new ways to cooperate and respond to constantly emerging new concerns about aviation safety. The history of this evolution and the efforts to implement change over time provide useful insights into how a complex aviation system fundamentally changed its oversight of aviation safety. This history offers insights that could also be useful to the Air Force inspection system and its role in sustaining and improving the Air Force aviation program over time. The following points are of particular interest:

- Throughout, the FAA appreciated the importance of culture and the importance of changing culture to successfully introduce qualitatively new ways of assuring aviation safety.
- To effect change, the FAA repeatedly applied the tools of formal change management, to include the development of a new consensus on a way ahead, defining a clear vision, providing extensive training, and updating data systems in all parts of the aviation system to support improved oversight.
- The implementation of such significant change has required decades.
- Congress and the FAA have taken advantage of "burning platform" opportunities to motivate change. For the FAA, these events that force change have been tragic aviation accidents. The very fact that major changes can so frequently be tied to these events suggests that change would not have come so easily without them.
- The use of voluntary self-reporting came relatively late in this process of change. This is true, in part, because it took time to realize the value of such reporting. But it also took time to learn how to protect self-reporters well enough to create a consensus for change that would allow for useful voluntary self-reporting.

These points appear in the following brief sketch of change over the last 50 years in the FAA.

The Creation of the Aviation Safety Reporting Program (ASRP)

Drawing upon the lessons learned from the World War II, the military and commercial aviation communities first recognized the need for a national incident reporting system during the 1958 FAA Enactment hearings. The FAA, however, did not formally enact a reporting system until after the TWA 514 crash of December 1974.[5] The TWA crash precipitated a comprehensive study of the U.S. air transportation system. As a result, the FAA formally established the ASRP in May 1975. The National Aeronautics and Space Administration (NASA), an independent agency outside the FAA, administered ASRP to better ensure confidentiality and objectivity of the data collection process and subsequent data analyses. To support its role, NASA implemented a formal data collection and assessment system, the Aviation Safety Reporting System, in April 1976. Over the years, as the ASRP proved itself to be effective, the concept gained industry-wide acceptance, and its implementation grew increasingly extensive.

Shift in Aviation Safety Culture in the Mid-1980s with the Post-Deregulation Traffic Surge

Deregulation of the commercial aviation industry in the late 1970s triggered a dramatic increase in air traffic and, along with it, in aviation safety concerns. In response, the FAA made a conscious effort to shift its approach to assuring aviation safety from one based on blame to a more collaborative, system-wide approach. The FAA began to lead this cultural shift in the mid-1980s. The International Civil Aviation Organization (IACO) and the global aviation community adopted it soon after.

The shift came to maturity in the United States when Admiral James B. Busey[6] became the FAA Administrator in 1989. Admiral Busey initiated the *Compliance for the '90s Program* (also referred to as the *C90s Program*). The C90s Program consisted of the following three key components, which rolled out in succession: (1) a clear vision, (2) training, (3) voluntary reporting systems.

Articulation and Communication of Clear Vision. First, Admiral Busey articulated a clear vision of moving away from a blame culture at FAA, which had a reputation for "shooting the wounded," toward a more collaborative, system-wide culture in which everyone worked toward a common set of goals. He demonstrated active endorsement and direct engagement by the FAA senior leadership by assigning his immediate reports—i.e., Associate Administrators—to communicate his vision. The Associate Administrators, together with the directors of line management personnel, conducted seminars across the country, targeting the PIs for operations, maintenance, avionics, and repair stations. These PIs were the key people in the FAA's inspection system. These seminars encouraged active dialogues and an interchange of ideas that would help to move beyond the "shooting the wounded" approach.

Key Personnel Training. Second, intensive training was targeted to PIs. It sought to (1) give them confidence that aviation safety could be assured even in their absence and (2) dissuade them from believing that heavy-handed enforcement by the legal community was the most cost-effective way to assure safety.

Establishment of Voluntary Reporting Programs. Third, in the early 1990s, the FAA initiated the ASAP, VDRP, and FOQA voluntary reporting programs described in Chapter Four. These programs encouraged collaboration, leveraged FAA resources so that safety could indeed be assured in their absence, and promoted a system-wide approach to solving safety problems.

---

[5]   All 85 passengers and seven crew members on the flight were killed from this crash. The accident investigation later found that, six weeks prior, a United flight had narrowly escaped the same fate.

[6]   Admiral Busey served in the U.S. Navy from 1952 to 1989 and was the Vice Chief of Naval Operations and Commander in Chief Allied Forces Southern Region of NATO. After retiring from the Navy, he became the FAA Administrator from 1989 to 1991 and, subsequently, served as the Deputy Secretary of the Department of Transportation.

One of the challenges of implementing the voluntary programs initially was resistance from the airlines about sharing the proprietary and confidential information collected from, for example, their employees through the ASAP and from in-flight data through the FOQA program. The carriers' primary concerns were uncontrolled disclosures of the proprietary data through the Freedom of Information Act and the potential litigations that can result from such disclosures. To respond to these concerns, the FAA was able to help in the passage of new legislation (14 CFR Part 193) in the mid-1990s, which established that ASAP, FOQA, and other FAA voluntary reporting programs were exempt from the FOIA. This legislation was one of the milestones that made the implementation of the voluntary reporting programs much more effective both from the FAA's and airlines' perspective.

Shift in Aviation Safety Culture in the Mid-1990s with Increased Airline Responsibility

The ValuJet 592 crash[7] in 1996 set the stage for another cultural shift. The crash instigated an intensive 90-day safety review of the accident, which revealed that ValuJet's business model, although approved by the FAA, was not well designed from a safety standpoint. The airline relied heavily on outside contract services for their operations—e.g., maintenance, baggage handling—and on other cost-cutting measures, which ultimately contributed to the accident. From the accident report, the FAA concluded that no amount of inspection could solve safety problems if the system was not well designed. The FAA hence initiated a shift toward a new approach to safety based on sound *safety system design* rather than *individual safety compliance* by individual actors in the aviation system.

This shift brought about a fundamental reexamination of (1) the definition of safety, (2) the roles of the FAA and the airlines in ensuring aviation safety, and (3) means to control safety risks. The FAA established the operational definition of safety as "an operationally acceptable risk level explicitly defined and maintained through continuous identification of safety hazards and continuous mitigation of those hazards and risks." New legislation assigned both the FAA and the air carriers direct responsibility for identifying safety hazards and continuously assessing the risks associated with these hazards. All safety regulations were interpreted to be risk control related; risk control mechanisms were considered to be integral to all aspects of commercial aviation safety and operations.

To implement this new approach, the FAA initiated ATOS (discussed in greater detail in Chapter Four), which, together with the various voluntary reporting programs, operationalized the new culture and specifically addressed the importance of safety system design principles, the active role of airlines, and the integration of a risk control and management approach to safety. ATOS has been refined continuously since its initial fielding in 1998[8] and has evolved into an effective and valuable tool used extensively by the FAA inspection community today.

To field ATOS, the FAA conducted an extensive outreach program in the late 1990s and early 2000s. The FAA (1) met with operationally critical leaders of every air carrier, like directors of operations, directors of maintenance, directors of safety, chief pilots, and chief inspec-

---

[7] All 110 people aboard the flight were killed from this crash.

[8] The FAA initially implemented ATOS in 1998 with the ten largest air carriers, which carried 98 percent of U.S. air passenger traffic. Currently, all U.S. air carriers participate in the system.

tors; (2) encouraged the airlines to use the ATOS structure for their own internal auditing/control processes (IAP);[9] and (3) provided intensive, formal training courses for all FAA PIs.

Significant Investment in Database Systems

The FAA and the industry together have made significant investments in developing major database systems and tools that formalized certification, data reporting/collection, and data analysis processes. These systems provided a formal, common, and standardized platform for the FAA and the operators to (1) automate their respective activities with significant efficiency gain and (2) share information based on common nomenclature.

As mentioned in Chapter Four, both ATOS and the voluntary reporting programs have been continuously improved and refined over the years since their initial implementation. At the time of this report, they are going through major overhauls and transitioning into next-generation systems. ATOS is in the process of transitioning into the Safety Assurance System (SAS); the plan is to integrate various voluntary reporting programs into a more comprehensive Safety Management System. Both SAS and SMS will have a more enterprise- and system-wide approach and will enable, where appropriate, the incorporation of national and system-wide trends into individual operators' safety programs and provide more direct linkage between SAS (inspection) and SMS (voluntary reporting).

Summary

Formal change management methods have been developed, applied, and refined in the U.S. over the last three decades. A consensus has emerged, in the academic and technical business literatures that study change management in the private and public sectors, on the factors that have helped ensure successful implementation of significant changes in large, complex organizations like the U.S. Air Force.

In general, it is helpful to conceive of a change as consisting of three elements: planning, execution, and sustainment. Planning sets the stage for a change by gathering senior leadership support, assembling a coalition of all stakeholders, articulating what the change is and how it helps the organization, and developing initial evidence that the change will produce its intended effects. Execution transforms the concept behind the change into concrete actions that include instituting the change throughout the organization, monitoring evidence of its performance, and adjusting the change to reflect experience to date. Sustainment transitions the change from its position of special oversight into full integration with the standing policies, practices, and processes of the organization. Large changes typically occur in increments that allow the leadership to monitor performance in a timely way and sustain support for ongoing change. As a result, planning, execution, and sustainment activities often occur simultaneously as each incremental change works through its full life-cycle action plan.

Throughout any change effort, a number of factors can affect the likelihood of success. Training ensures that the individuals who must change their own behavior for organizational change to succeed know how to effectively fulfill their new roles. The organization seeks to

---

[9] The airlines have their own set of internal auditing programs. For example, for their maintenance operations, airlines are required to have a Continuous Airworthy Maintenance Program (CAMP) to ensure that their aircraft are properly maintained.

motivate change in these individuals by giving them enough freedom—enabling them—to change their behaviors as required for organizational change. The organization then uses its normal instruments to motivate personnel, including rewarding individual behaviors that support organizational change and sanctioning those that do not. Testing and measurement monitors the status of the change as it proceeds. When such monitoring detects weaknesses or failures, the organization reacts quickly to mitigate the problems involved or to terminate a change that is not working as planned. When such monitoring confirms success, the organization cites this information as evidence to sustain the support of the senior leadership and the coalition of interested parties. Appropriate resource planning ensures that the organization makes resources available to support training, motivation, testing, and adjustment throughout the change effort.

The FAA's efforts from the 1970s to the present to change its inspection system help illustrate many of these ideas in practice. In a series of significant adjustments, the FAA gave special attention to:

- securing and demonstrating the ongoing support of its senior leadership
- cultivating ongoing efforts to create and sustain consensus among the government, private companies, and employee unions with significant stakes in FAA's regulatory mission
- encouraging extensive communication to keep all participants aligned to a common purpose
- training leaders and personnel about the meaning of each increment of change and how their roles would change
- allowing changes to start small and then expand as experience with and confidence in change grew
- investing in new data systems that would enable the participants to achieve the full potential of the changes
- shaping incentives to promote participation in change and adjusting these incentives as the evolving consensus among key stakeholders allowed for change.

Voluntary reporting systems played a central role in the FAA changes. These systems can yield the kind of data the FAA seeks in its inspection system only if personnel throughout the aviation system feel safe to report accurate information about current problems. The FAA has had to take extensive measures to induce honest reporting. It started with a concerted effort to shift the prevailing "shoot the wounded" culture among FAA inspectors to one more constructively focused on collaborative problem solving—an effort that conceivably resulted in greater psychological safety. The FAA then promised reporters anonymity, turned over administration of its data system to a third party to protect the data collected, gave reporters carefully defined exemptions from normal FAA sanctions for errors and failures, and even got Congress to change FOIA to protect proprietary data and data on bad outcomes in the aviation system.

Every major organization change effort has its own challenges and dynamics. The ongoing FAA experience offers insights about change that might inform Air Force efforts to effect an equally challenging set of changes in its own inspection system.

CHAPTER EIGHT

Recommendations

The results of our review of practices to emulate, investigation of Air Force personnel's experiences in the field, and literature review suggest recommendations for SAF/IG to consider as it develops and implements a new inspection system. The recommendations are organized in accordance with the following goals:

- general recommendations
- selecting a better inspection interval
- reducing the inspection footprint
- increasing the emphasis on self-inspections and self-reporting
- introducing the new UEI
- introducing MICT
- implementing significant organizational change
- conducting additional analysis to support implementation.

These recommendations were informed by ideas shared by Air Force personnel during our fieldwork, but they neither adopt all those personnel's suggestions nor are limited to inspector and inspectee remarks. Moreover, while we view them as constructive steps in the right direction, we cannot estimate the results of these changes or their cost-effectiveness without further analysis.

General Recommendations

Consider adopting a formal risk management system to guide Air Force inspection-related decisions and activities. Such a system would significantly enhance the Air Force's ability to make its inspection system more cost-effective. Cost-effectiveness is inherently about weighing costs and effectiveness. For the Air Force to do this well, it must define both clearly. A formal risk management system would help clarify the relative importance of different elements of a wing's compliance, readiness for a contingency mission, and execution of the current operational mission.

The FAA inspection system relies heavily on a risk management system that assesses such issues in three steps. In an Air Force setting, these steps would include the following:

- What is the inherent risk associated with each item that the inspection system addresses? What is the probability of a negative consequence if an inspection detects a deficiency? What is the magnitude of the negative consequence if it occurs? The system would identify items to be inspected that have higher probabilities of negative consequences than others as well as higher magnitudes of negative consequences when they occur.

- What additional or exceptional risks do a wing's specific operating conditions impose? Does the wing operate in an environment with unusually high probabilities and magnitudes of negative consequences? Does its own record point to higher risks of this kind? Do changes in its leadership, mission, or other factors unique to the wing present unusual risks?
- What resources does the Air Force have available during any period to inspect items subject to these risks? Given the resources available, where will inspection activities during any period mitigate risk the most?

In their efforts to improve the Air Force inspection system, SAF/IG and the ISITT have often asked analogous questions. Where is the risk the highest? Where are risks of injury, death, or mission failure easiest for the Air Force to accept or bear? Where should the Air Force place its constrained inspection resources? If budget constraints lead to cuts on those resources, where should the Air Force take the cuts? The FAA risk management system allows it to address such questions in a fairly well-informed manner. Until the Air Force develops an equally capable risk management system, it will be hard-pressed to generate well-informed answers to the questions above. Without them, it will be inherently limited in its ability to take full advantage of many of the changes it is pursuing. It will also be severely limited in taking advantage of a number of the recommendations we offer regarding the inspection interval and footprint below.

SAF/IG should take the lead in developing a risk management system suited to the new inspection system. The Air Force and FAA inspection systems differ in an important way. The FAA system focuses on safety, while the SAF/IG system addresses every aspect of Air Force performance. When the FAA developed its risk management system, it convened a group of subject matter experts with a relatively narrow range of expertise—issues relevant to safety in aircraft design, operation, maintenance, and support. In effect, the FAA took the substantive lead on creating the risk management system and continues to manage its own SMEs to adapt and refine the risk management system as the inspection system generates new information on the operation of the commercial aviation system. The SAF/IG community relies on functional area managers for comparable substantive input on the content of the checklists it uses to conduct inspections. As a result, it cannot create a risk management system by itself. Doing so would violate basic relationships among stakeholders relevant to the Air Force inspection system.

As the agent of the secretary and chief of staff responsible for the performance of the inspection system as a whole, SAF/IG could, in principle, lead an Air Force–wide effort to create a new risk management system. The ongoing effort to improve the inspection system could serve as a model, but (1) developing a risk management system would require coordination of far more detailed and technical information, and (2) sustaining a risk management system would require a long-term, Air Force–wide consensus on its importance and commitment to continue supporting it. SAF/IG could take responsibility for developing the basic structure of the system, including a template for identifying issues relevant to the inspection system and for assessing risks inherent in these issues and risks likely to vary across wings for each issue. SAF/IG could provide a long-term institutional home, perhaps at AFIA, for a program to apply this template regularly across the Air Force to keep the Air Force's official knowledge of the risks relevant to its inspection system up to date.

With suitable long-term support from the secretary and chief of staff, SAF/IG could lead such an effort, but it could neither create nor sustain it without significant support—in terms of leadership focus and resources—from the functional area communities that currently create checklists for the IG. A simple starting point for such an effort could be an Air Force–wide effort to attach risk assessments to each item on each functional area's checklist. Simply adding a requirement that any checklist item must be accompanied by a risk assessment could, in itself, impose discipline on the checklist items that functional communities identify.[1] To be useful, however, these assessments would have to be applied in a uniform way across the Air Force. That is not how checklists are created today. As a neutral honest broker, perhaps SAF/IG could take responsibility for structuring the conceptual approach to assessing risk and enforcing it over time. The guidelines on effective formal change management in Chapter Seven could aid the SAF/IG structure and sustain the Air Force–wide coalition that would be required to institute this structure over the long term.

Selecting a Better Inspection Interval

Look for ways to condition the frequency of inspection on risk management factors. The Air Force is near full agreement on holding a major wing inspection every two years, but it appears undecided about what that inspection might look like. The discussion in Chapter Two suggests that, if the Air Force had an appropriate risk management system (see above), cost-effective opportunities would likely exist to customize each major wing inspection to emphasize items on a schedule as risk dictates. Lower-risk items might be inspected only every four years. Higher-risk items might be inspected regularly every two years, with additional no-notice inspections if desired. Because the Air Force does not have a fully developed risk management system, it would be difficult to attempt this today. But if the Air Force begins to develop a risk management system, it could anticipate moving in this direction as one of the benefits of that investment.

Over the long term, revisit the decision to move to one major inspection every two years. FAA experience demonstrates the feasibility and value of using a broadly more flexible approach. Many Air Force personnel we talked to favor such an approach as well. If the Air Force had a risk management system comparable to that which the FAA maintains (see above), a flexible approach would allow the Air Force to match individual inspection events, including no-notice inspections, fairly precisely to information about inherent risk and a wing's demonstrated capabilities in the activities that these events examine. Moreover, a flexible approach would allow the Air Force to tailor each inspection to a wing's demonstrated capabilities at relatively low cost.

At the time of this report, the initiative to improve the inspection system appeared to be heading toward a two-year interval. As it moves to a two-year interval, the Air Force can keep a

---

[1] The Office of the Secretary of Defense is currently testing such an approach in its development of metrics to populate DoD's annual Government Performance and Results Act performance plan. The approach not only disciplines requests for metrics but also provides information on what information should be reported to reflect the real risks associated with each metric.

more flexible approach in mind and collect Air Force–specific data that would make it easier to revisit this decision in a more fully informed way (see discussion of pilots below). Alternatively, if the Air Force develops a better risk management system, it could anticipate migrating to a broadly more flexible inspection interval once the risk management system is mature enough to support such an approach.

Reducing the Inspection Footprint

As future external inspections are reduced in size and focused, ensure that they continue to capture the priorities of SAF/IG and the relevant functional area communities. Apparent redundancies among inspections, assessments, and evaluations occur close to one another point to substantial opportunities to go beyond synchronization and truly integrate as many of these events as possible into a single, more cost-effective oversight event. To do this, the Air Force will need to convene teams with IG and functional participation that can assess redundancy within and across checklists. As these teams do this, they should keep in mind that, even if an inspection and an assessment address exactly the same checklist item, they may emphasize different things when collecting information about this item. These teams must be prepared and empowered to weigh IG and functional priorities and ensure that they are appropriately balanced in the final checklists to address all of their needs. This balancing will be easier to justify and sustain if the Air Force can base the decisions of these teams on a formal risk management framework (see above).

Apply formal sampling guidance to reduce the burden of inspections and increase their productivity. A variety of sampling approaches, some more scientific or deliberate than others, pervades the inspection system today, affecting what portion of Air Force and MAJCOM priorities are in fact addressed in any particular inspection, how much confidence the Air Force should have in the items that are addressed in a wing inspection, and what kinds of behaviors inspections drive at the wings they visit. More formal sampling will be feasible only if the Air Force develops a more formal approach to risk management in its inspection system (see above). If it does, the Air Force could use that risk management system to align the checklists that the Air Force as a whole and individual MAJCOMs create with what is feasible to measure at the wing level with the resources available. The FAA has extensive experience doing this that the Air Force could learn from. The Air Force could also then give its inspectors more formal training in sampling strategies and use its risk management system in help individual inspectors apply these strategies. Finally, the Air Force could use its risk management system to clarify what local behaviors it wants to drive and translate these goals into explicitly designed local sampling strategies. RAND's work on designing performance-based accountability systems offers guidance that the Air Force can apply here (Stecher et al., 2010).

Use information on a wing's past performance to design the focus and depth of a full inspection. The Air Force can achieve its goal of ensuring that every wing leadership team faces at least one major inspection during its tour by scheduling at least one major inspection every two years. Once that goal is achieved, the Air Force can then turn to the question of how much effort to place into such an inspection when it occurs. Presumably, if strong evidence exists that a wing is complying appropriately, is ready for assigned contingency missions, and is executing its current mission well, a scheduled biennial major inspection need not be as demanding as it would be if the wing displayed weaknesses. The IG can draw on data available from past inspection events and from other monitoring systems like readiness reporting, AFMA's Climate Survey, AFSC's AFCAST survey, and regular data feeds from a wing's self-inspection program (and, ultimately, the CCIP) to gather information on the wing's likely status. The IG can also use small no-notice inspections to probe particular aspects of the wing's operations based on what existing data systems suggest. The IG can then tailor each biennial inspection to anticipated conditions before the IG inspectors and their augmenters arrive at any wing. The better the Air Force's risk management system (see above), the greater the pay-off will be of tailoring major inspections in this way.

Increasing the Emphasis on Self-Inspections and Self-Reporting

Consider adapting some aspects of the FAA's voluntary reporting system as part of the new CCIP. The FAA's successful use of a system of systems, including enterprise-wide programs as well as ones focused on specific elements of its organization, offers a number of practices that the Air Force could adapt for its own use. For example, careful protection of the anonymity of voluntary reporters has given the FAA an independent channel of information that it can use to double-check information flowing through more traditional channels that reporters might attempt to distort to their own advantage. The Air Force could emulate this directly in a variety of ways (see below). In addition, the Air Force could employ SMEs to analyze instances of non-compliance across wings, and then use the SMEs' findings to serve as the basis for Air Force–wide or MAJCOM-wide watch lists, alerts, and suggestions for corrective actions.

Foster conditions for psychological safety. Both research and practice indicate that psychological safety can influence voluntary, proactive error reporting. Wing leadership can engage in efforts that foster psychological safety, such as conveying a message to personnel that committing errors may be permissible under certain circumstances, rewarding error reporting, and holding individuals accountable for "blameworthy" acts. Implementing "blameless reporting," a system in which the reporter of a deficiency can do so without fear of reprimand, is another important step. Along those lines, the FAA has taken extensive measures to create a sense of safety, measures that offer concrete examples for SAF/IG to consider. It started with a concerted effort to shift the prevailing "shoot the wounded" culture among FAA inspectors to one more constructively focused on collaborative problem solving. The FAA also promised reporters anonymity, turned over administration of its data system to NASA to protect the data collected, gave reporters carefully defined exemptions from normal FAA sanctions for errors and failures, and even got Congress to change FOIA to protect proprietary data and data on negative outcomes in the aviation system.

Support wings' efforts to preserve the external look. During our fieldwork, we learned that either in preparation for an inspection or as part of their self-inspection program, wings frequently obtained an outside perspective to identify areas in need of improvement. This fresh set of eyes was sometimes from another part of the wing, sometimes from a separate wing, and occasionally from higher headquarters staff, but in every instance, a neutral assessment was provided that personnel believed helped the wing's self-improvement efforts in ways that additional internal scrutiny could not. Such was the case even if the outside parties were lacking in functional expertise or specific training in inspections; in those instances, the questions they asked to educate themselves gave wing personnel an opportunity to explain processes and show evidence to their visitors. Accordingly, as SAF/IG integrates external inspections and assessments and decreases either the frequency or scope of IG-led inspection events, it should also encourage, via policy or other guidance, the use of external looks by non-IG personnel who can take an unbiased look at a wing's practices.

Introducing the New Unit Effectiveness Inspection

Recognize that leadership and discipline are multi-faceted constructs and measure them as such. Views expressed by Air Force personnel during our fieldwork indicate that discipline has multiple facets: customs and courtesy, adherence to rules, and possibly legal or fitness-related elements. Leadership was not only discussed during our interviews and focus groups as a sum of many attributes, but has been measured in scholarly literature and existing Air Force survey instruments in this manner. As SAF/IG moves forward with plans to formally measure discipline and leadership within the new UEI, it should adopt multiple measures, both inter-related and independent, quantitative and qualitative, to obtain a robust picture of the status of leadership and discipline within a wing. Along those lines, indicators of poor or insufficient leadership and a lack of discipline may be as telling as measures of good leadership and strong discipline.

Ensure measures of leadership take into consideration the impact of the full chain of command, not just the wing commander. While a wing commander sets the tone for his wing in a number of ways, his span of control and direct interaction with many wing personnel is intentionally limited. Leadership at the group, squadron, and flight levels, including both officers and NCOs, contributes to the overall functioning of the wing and, accordingly, should be included in IG assessments of the leadership climate. The use of statistical "data aggregation" techniques and survey items about the chain of command, such as those included in AFMA's Climate Survey, are two ways of accomplishing this.

Use existing data to inform the inspection process. Inspectors should make use of data collected for other purposes to guide sampling decisions and corroborate information they themselves collect during the UEI. For example, the results of an AFCAST survey can suggest areas that merit a closer look during an inspection, and data such as fitness ratings or the nature and extent of various legal infractions could complement the IG's own assessment of unit discipline. As the UEI is developed and gradually implemented, inspectors could also be on the lookout for other naturally occurring data sources or systematic data collection efforts already in place (e.g., Inspector General Deputy Director, Complaints Resolution Directorate [IGQ] data). Making use of existing data sources is not only cost-effective, but the extent to which various sources point to the same conclusion can increase confidence in the validity of final inspection results.

Develop a new UEI survey that adopts items from existing survey instruments. Under the United States Code Title 10, Sec 8020, IG organizations are able to administer surveys in the execution of their responsibilities. Moreover, AFI 38-501, Air Force Survey Program, not only notes that IG organizations are exempt from the control measures generally in place for systematic data collection within the Air Force, but also states that IG organizations can use other aspects of the Air Force Survey Program (e.g., survey development, recruitment) as appropriate. Accordingly, SAF/IG has a strong basis to develop and implement its own survey as part of the UEI. This survey could make use of survey items from scholarly literature, such as the MLQ's measures of transactional and transformational leadership and Edmondson's

(1999) measures of psychological safety, which have been validated as methodologically sound and applicable in diverse settings. Similarly, items from the Climate Survey and AFCAST set of surveys could be used by the IG in its own survey, either in addition to or instead of relying on the data collected from and reports generated by AFMA and AFSC. This may be particularly helpful if, due to confidentiality considerations, only data at a high level of aggregation are available to the IG.

Consider the use of qualitative measures, but ensure they are standardized across inspection teams and sites. Interviews can yield useful insights, such as helping to explain the results of a cross-sectional (i.e., one-time) survey, and observations may be the best way to obtain information about certain aspects of unit functioning (e.g., task evaluations, interactions between leadership and personnel). However, measuring a different set of indicators at each wing or measuring the same set of indicators differently across wings will render comparisons of wing performance within the same MAJCOM or across the Air Force difficult, if not impossible. Similarly, trend analysis of a wing's performance over time is compromised by variation in the methods used to assess compliance, readiness, effectiveness, and other outcomes deemed critical by the Air Force. While surveys are an easy way to quantify concepts and measure them consistently across settings and time, they are not the only way to do so. For example, checklists of specific observations that inspectors should consistently make while on site at a wing could be developed, along with guidance about the type of documentation (e.g., photos, tallies, narrative) that should accompany each observation. Even qualitative indicators such as whether personnel consistently stand up when an officer enters the room, can be quantified, for instance, and indicators that do not lend themselves to yes/no measurement can be documented in a consistent manner if specific guidance is given about what to include in narrative parts of an inspection report. Qualitative measures can indeed be analyzed in a way that facilitates comparisons across settings and across time, but obtaining those measures in a uniform way is a critical prerequisite.

Introducing the Management Internal Control Toolset

Follow through to ensure that MICT is implemented effectively. By many accounts, MICT appears to offer well-designed software. But, based on their past experience with other faulty implementations of information technology, Air Force personnel in the field are skeptical about the future usefulness of a standard wing system like MICT.

Standard lessons learned from the successful introduction of information technology in the past should be helpful here: Provide training designed to address users' needs. Train personnel when the system is introduced and maintain ongoing training programs to ensure that a sufficient number of relevant personnel in wings remain trained as personnel turn over or are deployed. Maintain a real-time help line with a human at the other end of the line. Update the system as experience with the new system reveals opportunities, but keep changes simple, so that they do not precipitate a need for significant additional training. To help with this, ensure that changes are planned and approved by a users' group and based on user needs, rather than planned and approved solely by information technology specialists. Sustain the new system by resourcing these efforts as a planned cost of doing business.

Recognize MICT as a complement to external inspections and assessments and internal self-inspection, not a replacement for them. The increasingly constrained resource environment that the Air Force is likely to face raises special concerns about the costs incurred by MICT. Experience to date suggests that, on net, MICT actually tends to free wing resources by simplifying the entry and maintenance of data enough to more than cover the time required for additional training. To the extent that this proves to be true as MICT spreads through the active component, MICT may also free enough resources to cover the additional time wing personnel spend using MICT to track and finally close corrective action plans. Wings should not assume that MICT can displace normal face-to-face oversight activities. The Air Force should create and enforce policy that ensures that wings do not use MICT to attempt to "save" resources in this way. If the Air Force ultimately finds that it must program additional resources to sustain MICT, it may be time to seek an alternative. Only time and close monitoring will tell.

Implement and sustain an approach to using MICT that maintains (1) standard core information and (2) wing-unique information. The core information should be tied to a standard set of checklist items of interest to a MAJCOM or the Air Force as a whole. This will allow expeditious external access to data in MICT, if that is permitted, and reliable auditing of the information of greatest interest to the MAJCOMs and the Air Force as a whole.

Wing-unique information can address issues that only one wing faces or issues that any wing, over time, has come to believe deserve additional attention. Today, wings routinely develop their own information management systems to manage and track such information. If MICT cannot accommodate such information, wings will maintain legacy systems side by side with MICT, potentially reducing its cost-effectiveness by compounding the costs of sustaining training, help-desks, updates, and other support for multiple systems.

Maintain ways to double-check any information in MICT that is freely available to external overseers at the MAJCOM or Air Force level. A major goal of MICT has been to let external observers look over the shoulders of personnel in a wing to see how they are doing. Extensive experience in performance accountability systems reveals that the quality of data of this kind can easily be compromised by individuals who do not want external observers to see how they are actually doing.[2] SAF/IG can double-check this information in multiple ways. It can intermittently conduct stringent audits of numbers reported, especially in units with a past history of faulty reporting. It can use voluntary reporting systems that guarantee reporters anonymity or effective amnesty (we discuss such systems elsewhere in this chapter). Finally, it can compare data reports from different sources in a wing to ensure that they are internally consistent.

Experience also suggests that SAF/IG should be prepared to adjust the data it requests when it finds that the requests themselves are inducing behavior within wings that allows honest reporting but compromises the effective performance of the wing. In practice, an ongoing "cat and mouse" game between overseer and overseen is not uncommon. Inspectors must remain vigilant and adaptable to play effectively in this game.

Implementing Significant Organizational Change

As change in the inspection system goes forward, keep in mind that it has many moving parts and operates as part of a broader governance system. The design of the inspection

---

[2]   For more detail on how to do the things proposed in this recommendation, see Stecher et al., 2010.

system depends on the design of all its parts. For example, decisions about appropriate inspection frequency, inspection footprint, and self-inspection regime all depend on one another. As the inspection system changes, such elements of the system will probably have to change in sync to ensure that the governance system as a whole continues to function well.

In particular, the better motivated and capable the leaders and airmen in a wing are to promote Air Force–wide goals, the less work the formal inspection system has to do. Those goals might involve compliance, readiness for contingency missions, or execution of the current operational mission. The inspection system should be designed and prepared to become more intrusive in wings that are demonstrably not pursuing such Air Force–wide goals appropriately or effectively.

Elements of the currently anticipated new inspection system involve changes that are designed to improve the motivation and capability of the leaders and airmen in wings. For example, wings will have better self-inspection systems, will have better data management systems, and will routinely report more complete information to MAJCOM IGs and functional area managers than they do today. Aligning the inspection schedule to the tour length of the wing leadership may increase the motivation of wing leaders to give compliance and readiness for contingency missions greater attention.

If these changes have the desired effects, they should allow the Air Force to reduce its external inspection of wings. But the Air Force inspection system should not just assume this will occur. Instead, it should monitor changes closely and verify that they are having their desired effects before withdrawing external inspections. This could be difficult, because change is likely to be much more effective in some wings than in others, and the Air Force's current and planned inspection systems are not well suited to treat separate wings differently. Keeping all parts of the governance structure properly balanced as change in the inspection system goes forward will be challenging.

Full implementation will take time. The Air Force should plan for this. Changes of the magnitude that the Air Force is currently considering take many years. Comparable changes in the FAA inspection system occurred over decades. The FAA changes were implemented slowly, in part, because it took time to build consensus among the key stakeholders to support critical elements of change. There were also many moving parts that had to be adjusted and synchronized as experience accumulated and the FAA and its partners learned how to structure the new system to their mutual advantage. The challenges that the Air Force faces may not be as daunting, but they are significant.

A common response to the demand for time to implement change is that the senior leadership of the Air Force turns over too often to sustain and coordinate changes that take longer to implement than the time any one leadership team is in place. But the Air Force routinely develops and sustains weapon systems over decades, calling on many successive leadership teams to share custody of a single system over its lifetime. The Air Force can similarly handle a significant change in its inspection system by breaking it into incremental chunks, monitoring each chunk, letting each team take credit for the progress of the chunks it oversees, adjusting each new chunk on the basis of what was learned through the last one, and sustaining a coalition over time that supports this incremental approach.

Use formal pilot tests to help monitor and refine increments of change before they are implemented throughout the Air Force. Pilots offer a chance to try a new idea through limited application, in which unanticipated weaknesses can be corrected before they cause significant harm. They are not simple demonstrations. Rather, they identify relevant metrics of performance and resource use, baseline these metrics before the start, clearly define and freeze a specific change for a fixed period of time, measure the resulting change using performance and resource use metrics, and adjust these metrics for any effects from factors beyond the control of pilot managers. They are quick and may not be totally rigorous. But they are systematic, objective, transparent, and clearly documented.

The new inspection system being developed offers many opportunities for pilots. For example, MICT was introduced in the reserve component before it was considered for use Air Force wide. Introductions in selected reserve wings could have been instrumented and monitored so that the Air Force had ex post measures of the effects of the change that it could use to adjust MICT before expanding its application. This could still be done within selected reserve component wings where MICT has been in place for a while, or it could be done within certain active component wings selected to "lead the fleet." Traditional "lead the fleet" exercises in the Air Force closely instrument a small number of aircraft to understand their performance in enough detail to be able to project how the rest of the aircraft in the Air Force will perform under similar circumstances.

Many other pilot opportunities exist; M-ASAP is one. AMC has already been using M-ASAP, so it is too late to structure its introduction as a formal pilot. But the Air Force could begin to collect systematic data on AMC's experience with M-ASAP to determine (1) whether it warrants broader application in the Air Force and, (2) if so, how it should be adjusted before its application is expanded. Initial efforts to develop risk assessments of checklist items could easily by piloted by function and/or MAJCOM. The use of no-notice inspections to tailor a follow-on major inspection could be piloted by function at the wing level. Formal programs to foster psychological safety in wings could be piloted in individual wings. Application of data from AFCAST or the Climate Survey could be piloted in individual wings. Measures of leadership attributes could also be piloted in individual wings.

Anticipate and disarm negative perceptions about proposed changes. Among the most important moving parts that must change as the Air Force implements a new inspection system are the members of the Air Force—military, government civilian, and contractor—who will inspect and/or be inspected. Our expert interviews and interactions with Air Force personnel in the field revealed many good ideas about potential change, but also a general lack of understanding of where the Air Force is currently heading and how all its parts must change to make the proposed changes viable. During our fieldwork in particular, both inspectors and inspectees expressed concerns about changes to the inspection system. Inspectors were skeptical, for instance, about placing greater emphasis on wings' self-inspection programs because (1) their experience to date with SIPs has been poor and (2) they are unaware how the new system will improve the value of self-reporting. Inspectors also challenged the idea of credibly measuring leadership within the context of compliance inspection, perhaps because they have never seen leadership measured before as part of a standard compliance inspection. In addition, personnel from recently inspected wings often reacted negatively to plans to implement MICT, in part because of recent experience with new information systems that fell short of the promises made about them and imposed greater costs than they relieved.

These personnel must change their individual behavior to ensure the success of the changes the Air Force is seeking in its inspection system. Their views reinforce the premise, discussed in Chapter Seven, that successful change requires a compelling explanation for why it is necessary and how it will make things easier moving forward. This message should have a strong factual basis, but also needs to make an emotional appeal to personnel. In addition, if skepticism or negative views are based on misinformation, providing accurate information to correct these perceptions can help to overcome psychological barriers to change. Similarly, if concerns stem from perceived shortcomings of the current system, clearly explaining how the new system differs from and improves upon the current system can encourage a move away from the status quo toward a new, improved inspection system.

Accordingly, the Air Force should develop strategies that leaders throughout the chain of command can use to address negative perceptions of inspection system changes and refine them as needed throughout the execution and sustainment phases. For example, the Air Force can collect information on the effects of ongoing change and use that information to demonstrate the benefits of change—to the leadership, to those elsewhere who must change their behavior, and to the Air Force as a whole. These techniques, in turn, can give individuals the resources they need to support change, the freedom to change their behavior in positive ways, the skills to respond in ways that benefit the Air Force, and feedback that makes it clear how well each of them is doing at supporting change.

Conducting Additional Analysis to Support Implementation

Continue formal analysis of selected issues where full implementation will benefit from more technical depth. During the course of our analysis, we identified a number of topics that we would have pursued if adequate time and resources were available in the context of this project. The Air Force could pursue these as it continues to implement the inspection system changes already under way. They include the following:

Develop more detailed and quantitative analysis of the costs of the inspection system. The data currently available to the Air Force are not adequate to support the kinds of careful cost-benefit analysis it will need to refine its new vision of the inspection system as implementation proceeds. This task would address the costs imposed on units by external inspectors, the costs of self-inspections, and the costs to the external IG and functional communities of conducting external inspections. To the full extent possible, it should seek to state these costs in comparable terms—for example, person days or dollars. It should also seek cost drivers in each of these three areas that are relevant to decisions that the Air Force will need to make as implementation of the new inspection system proceeds.

Translate the risk assessment *framework* recommended here into guidance for an Air Force risk assessment *system*. Just as the FAA inspection system has evolved a risk assessment system tailored to the industry setting in which it operates, the Air Force could develop a risk assessment system tailored to its inspection priorities, culture, and organizational structure. Such a system would identify specific roles and responsibilities throughout the Air Force; specific requirements for data, data management, and the analytic capabilities that the Air Force would use to apply data; and specific goals for positive and negative incentives associated with data collection as well as institutional designs to instantiate these incentives. The FAA experience teaches us the importance of developing such a system in an evolutionary way that learns over time. This task would provide enough concrete detail to initiate this development process and offer guidance on how to sustain learning as the process proceeds.

Develop concrete and specific guidance that translates formal sampling methods into instructions that inspectors could apply in practical ways to increase the quality of information they can collect with given resources. The statistical, engineering, and social science research communities have access to extensive tools that they use routinely to sample data when they conduct empirical analysis. The Air Force inspection community needs access to such tools in a form that its inspectors can use. Such analysis would synopsize the formal concepts relevant to Air Force inspections, assess which are most easily applied in a setting with real inspectors, and draft explicit guidance for application in the Air Force inspection system.

Develop the basis for a more precise and operational definition of discipline. In the context of this analysis, the definition of discipline would focus on attributes that could be monitored and assessed in a compliance or readiness inspection. But an established, operational definition would likely be useful in a broader context across the Air Force. The more clearly a definition can support an empirically detectable link between unit discipline today and organizational performance tomorrow, the more influential the definition is likely to be in any inspection that monitors and assesses it.

Translate the current, broad implementation guidance offered in Chapter Seven into a form more tailored to an Air Force setting. Ideally, the first part of such a task would develop a detailed implementation plan with guidance on how to administer it and update it in response to accumulating evidence on the success of implementation. The second part would support the Air Force in its ongoing implementation and draw lessons learned for future major Air Force inspection program changes.

Analysis of Practices the Air Force Inspection System Might Emulate

SAF/IG asked PAF to identify practices that might help improve the performance of the Air Force inspection system.[1] A good place to look for useful ideas is within an organization whose peers perceive it as "best in class" with regard the kind of practice in question. So, because its peers thought for many years that Toyota used the most cost-effective approach to automobile design and manufacturing in the world, its Toyota Production System, also known as "lean production," became known as a best practice and even became an integral part of Lean Six Sigma, the current toolset of choice for improving complex processes. An organization like Toyota documents how its performance improved by applying the practice. When its peers test such claims, they learn over time that they can improve their own performance by emulating the practices of this "best in class" organization. There is circularity in such perceptions, but evidence that the practice yields demonstrable improvements in a real setting ultimately breaks the cycle, at least in those settings where organizations have successfully adapted the practice to their own needs and priorities.

From this perspective, the FAA has demonstrated dramatic improvements in aviation safety since it began to adopt and refine the approach to aviation system inspection that it uses today. There is preliminary evidence that measures of the level of quality in the safety climate that AFCAST measures can be correlated to broader measures of performance in the organizations to which it and its close analogs have been applied. These are, therefore, reasonable places to look for useful ideas.

One common analytic practice based on this principle is to anticipate that, if an organization makes its operating procedures more "rational" by applying formal analytic methods to sustain and refine those procedures over time, its performance will be better than if it does not.[2] In effect, this is ultimately the argument that underlies broad global support for International Organization for Standardization (ISO) 9000 management standards.[3] Buyers now generally prefer suppliers certified by ISO 9000 to those that are not because they believe such a certification carries useful information. The FAA's general approach to inspection applies many of the principles of the ISO 9000 family of standards. More specifically, its extensive application of formal risk assessment can be seen as evidence that the FAA documents inspec-

---

[1] For more information on how to recognize and access practices worth emulating, see Camm, 2003.

[2] For a discussion of this definition of "rational" and its implications for whether it is rational—in their own best interests—for organizations to structure themselves rationally, see March, 1994.

[3] For a discussion of the implications of ISO 9000 standards for process improvement in a defense setting, see Camm et al., 2001.

tion processes precisely and monitors their application to ensure that they mitigate risk cost-effectively. Similarly, AFCAST's increasing alignment to the DoD-wide taxonomy for classifying risk, the Human Factors Analysis and Classification System (HFACS), can be interpreted as similar evidence of a "rational" process in that it uses a formal rational tool that has helped improve outcomes in other settings.

At the end of the day, however, it is hard to accept that any practice is worth emulating unless it behaves in a way that is likely to add value to an organization. If we can show how a practice adds value in its current setting and how it might add value in an alternative setting—like the Air Force inspection system—it is easier to imagine that the practice is worth studying and adapting. Economists commonly apply such reasoning, believing that, if a practice survives in a highly competitive setting, it probably does so only because it adds enough value to compete successfully with potential alternatives. We apply an approach very similar to this in Chapters Two and Three to posit that, if an organization were to design its governance structure explicitly to improve its own performance, it would structure its inspection system as an integral part of that governance structure. We can use this approach to support an argument that the FAA's inspection system is designed to do exactly that, and to ask why this approach might increase value in the FAA setting, but not in the Air Force setting.

In a government setting, a tension often arises between commercial and government practices worthy of emulating. Because the global commercial setting is so much more competitive and so much larger than the U.S. federal government setting, it is far more likely that we will find practices worthy of emulating in the commercial than in the federal government setting. But it may be that the commercial setting is so different from the government setting that it is better to look for ideas closer to home—ideas that will be easier to adapt to another government setting, even if they do not offer as much potential for improvement as a commercial practice might. Such an argument offers a good reason to look to the Air Force Climate Survey for ideas on how to measure leadership. If the Air Force inspection system seeks better ways to measure leadership, it is not asking *whether* to measure leadership, rather what measures best capture what the Air Force values about leadership. Air Force Climate Surveys have been measuring unit leadership, down to the squadron level and below, for many years. During this period, myriad Air Force leadership teams have had a chance to refine the definitions of leadership that they value. The measures in the Climate Survey have, in effect, survived repeated tests of the Air Force leadership itself. AFCAST offers measures of leadership as well, but, as a relatively new system, it has not been tested in the same way. Also, it measures leadership from a perspective—aviation safety—that is narrower than that relevant to the Climate Survey and the Air Force inspection system.

For these reasons, we have turned to the FAA inspection and risk assessment systems and to the Air Force's AFCAST and Climate Survey systems for information on practices that the Air Force inspection system might emulate. Each chapter and appendix in which we discuss these systems offers additional information on how best to interpret practices relevant to the systems and adapt them to the Air Force inspection system.

- FAA: We interviewed managers of certain segments of the FAA inspection and risk assessment systems that appeared to be most relevant. We reviewed the extensive regulations and support documents relevant to these systems. We gave special attention to the FAA's oversight of air carriers, which account for about 80 percent of the FAA's oversight activity. Within this scope, we gave special attention to how the FAA uses formal risk assessment to choose inspection intervals, seeking to understand the implications of the FAA approach for the Air Force inspection system.

- AFCAST: SAF/IG asked us to give special attention to AFCAST. We interviewed the analysts who developed the first versions of it at the Naval Postgraduate School and the personnel at the Air Force Safety Center who currently administer and refine the AFCAST surveys. We reviewed the analytic literature on which it was built, the analytic literature on its application in many different settings (including those outside the safety domain), the efforts made to continue to link it to two different underlying safety-centric analytic frameworks, and its ability to explain variations in safety outcomes in different kinds of organizations.

- Climate Survey: We came late to this survey. We collected past versions of the survey and reviewed Air Force literature on its application in the past. We also discussed its structure with RAND analysts who have previously studied the use of Climate Surveys in a personnel management setting. If additional resources were available, we would collect more up-to-date information about the Climate Survey, but are satisfied that the general qualitative conclusions we draw here are valid.

Analysis of the Experiences of Air Force Personnel in the Field

Our fieldwork consisted of three efforts: (1) observation of a wing-level compliance inspection, (2) focus groups with inspectors, and (3) interviews and focus groups with members of recently inspected wings (i.e., "inspectees"). We worked closely with SAF/IG to decide which MAJCOMs to study, including which wings to visit during the period available for our fieldwork, April through July 2011. We mutually chose three MAJCOMs: ACC, AMC, and AETC. ACC and AMC are the largest and most important operational MAJCOMs in the Air Force. AETC is an important support command that also conducts aviation activities similar to those of ACC and AMC. Discussions in the ISITT repeatedly revealed how different potential changes at active component wings were likely to be from those at reserve component wings. With SAF/IG's agreement, we focused on active component wings to achieve as much depth as possible with the limited resources we had to apply to fieldwork.

Ultimately, this approach resulted in observation of a compliance inspection at a AMC wing; visits to recently inspected wings in AMC, ACC, and AETC; and focus groups with inspection team personnel at AMC, ACC, and AETC MAJCOM headquarters. The compliance inspection was intended largely to help us at RAND understand how the compliance inspection process described in AFI 90-201 is actually conducted, and to suggest topics to explore further via interviews and focus groups. As such, we took notes and shared them among project team members, but we performed no formal, structured primary data collection design and analysis. The remainder of this appendix accordingly focuses on the primary data collections that we did structure and analyze formally—the interview and focus group portions of our fieldwork.

Sample Recruitment and Composition

Inspectors

Inspector focus groups were conducted with IG core personnel, and, as applicable, with functional augmentees serving in staff positions at the time of our research. These sessions were conducted during visits to AMC, ACC, and AETC headquarters at Scott, Langley, and Randolph Air Force Bases, respectively. At each location, a local point of contact (POC) was identified to assist the project team in scheduling focus group sessions. The POC identified a date that would work for as many personnel as possible given the inspection schedule, planned leave, and other events (such as changes in command) and he or she worked with the RAND team to populate the focus groups as well. Given how inspection teams operate during an actual inspection, we opted to have a mix of officers and enlisted personnel participating in the same focus group sessions. In some sessions, civilian employees (often retired military with IG experience) participated as well. Although the focus groups were not based on a random sample and their results are not generalizable to the entire IG workforce, a broad range of experiences was represented in the sessions. In total, we conducted nine 90-minute focus groups, three per MAJCOM. The number of attendees per session ranged from five to 11 inspectors. In total, 71 individuals participated.

Inspectees

In order to ensure their inspection experience was highly salient for Air Force personnel, we sought to visit each recently inspected wing soon after its compliance inspection. We visited two wings one to two weeks after their compliance inspections, and the third wing was visited six weeks after its inspection. As was the case with inspector focus groups at MAJCOM headquarters, we worked with a local POC to identify a date convenient for the wing and to populate our interviews and focus group sessions with the desired mix of wing leadership and personnel.

At each wing, we requested 75-minute interviews with the leadership element (commander, deputy or vice commander, and chief) of the wing and of the operations, maintenance, and mission support groups (or equivalents). While we did interview representatives from wing and group leadership for all three recently inspected wings, scheduling constraints occasionally limited the amount of time we had with leadership or the actual officers in attendance for the interview. In total, we conducted 12 leadership interviews, four per wing, in which 27 people participated (one to four per interview).

We also scheduled 90-minute focus groups with officers, NCOs, and senior NCOs at each wing. In contrast with the inspector focus groups, the focus groups for the inspected wings were stratified by pay grade; separate sessions were held with officers (O-2 to O-4), NCOs (E-5 to E-6), and senior NCOs (E-7 to E-9). By design, focus group participants represented a variety of groups and functional areas (e.g., maintenance, health services, personnel, intelligence), ensuring that even though this was not a random sample, diverse perspectives would be represented. Overall, 69 personnel participated in nine focus groups (three per wing), which ranged in size from four to 11 individuals.

Fieldwork Topics

During the interviews and both types of focus groups (inspector and inspectee), we covered the following topics:

- inspection preparation
- inspection process
- potential inspection system change: reducing the inspection footprint
- potential inspection system change: greater emphasis on unit self-inspection
- potential inspection system change: use of MICT as an enabler
- leadership
- discipline.

Some of the actual questions varied depending on whether we were speaking with inspectors or inspectees. For example, during our sessions with inspectors, we discussed sampling within the context of inspection preparation, while with members of recently inspected wings, we asked how far in advance of the inspection preparation began and what it entailed. Within the inspectee sessions, we also asked questions about self-inspection program characteristics and error/non-compliance reporting. The interview and focus group protocols we used are provided below.

Analysis

All interviews and focus groups were led by a RAND researcher. In every session, one or two additional RAND researchers served as dedicated note takers, using participants' own words as much as possible in their documentation. After notes were drafted, the moderator reviewed and finalized each set of notes, which were then analyzed using QSR NVivo 9. NVivo 9 is a software package that enables its users to review, categorize, and analyze qualitative data such as text, visual images, and audio recordings. Software like NVivo 9 permits analysts to assign codes to passages of text and later retrieve passages of similarly coded text within and across documents. NVivo 9 is also capable of simple word-based searches as well as more sophisticated text searches, such as Boolean searches involving combinations of codes.

Our project team developed coding "trees" to facilitate the tagging of relevant interview excerpts. A coding tree is a set of codes that serve as "labels for assigning units of meaning to information compiled during a study" (Miles and Huberman, 1994, p. 56). Codes are used in the data reduction process to retrieve and organize qualitative data by topic and other characteristics. For this effort, codes were largely based on the interview and focus group protocols (e.g., inspection preparation, sampling, error reporting),[1] and three members of the project team worked to review and code notes from the 12 interviews and 18 focus groups. An iterative process of coding a few sets of notes, sharing examples of coding, and making refinements as needed was used to ensure the original set of codes was applied to text in a suitable manner. The coding of passages related to inspection preparation provides a good example of this approach: For inspectors, we first coded passages related to any aspect of inspection preparation. Then, after review of those passages and discussion among our team members, we developed and applied more nuanced codes pertaining to inspection preparation data sources, the extent to which inspection events are planned in advance, and sampling. Similarly, passages related to leadership and discipline were later broken down into those that discussed the definition of leadership and discipline, the measurement of each concept, and the perceived relationship between each concept and unit outcomes like performance.

After all the interviews and focus groups were coded, coding reports were generated to provide the frequency with which each code was applied. Additional reports were produced so that all the passages under a specific code could be reviewed together. Analyses of these reports were then considered in conjunction with other data sources (e.g., best practices analysis, literature reviews) to answer research questions key to the project's goals.

---

[1]   Full coding trees are available by request from the authors.

Inspection Team Focus Group Protocol

1. We're going to begin by going around the room. Please each take a turn and tell us your current pay grade, how long you've been in the Air Force, and your experience on IG-led inspection teams, including your typical role on those teams and how many compliance inspections you've helped conduct.

   [If roles are varied, instruct participant to focus on role(s) served in compliance inspections.]

   Thanks for providing that information. Now that I know something about your individual backgrounds, I'll move into the first set of questions for group discussion. They pertain to how inspectors prepare for a unit visit.

2. First, does an inspection team typically research a unit before visiting it for a compliance inspection, or does it deliberately go into an inspection blind, without advance information about the unit?
   a. *If it learns about unit pre-inspection:* How does an inspection team learn about a unit before its visit? What information sources are used?
   b. *Probe:* What about historical data from the unit, like performance or training records?

3. How much of the compliance inspection is planned in advance? Put another way, to what extent does a team know what it's going to look at and whom it's going to speak to before arriving onsite?
4. We understand there are situations in which an inspection team might use its discretion to inspect a subset of activities or functional areas rather than 100% of them. In that situation, how does a team choose what to prioritize?
   a. *Prompt as needed:* What were your teams' selection criteria?

   Our next set of questions is about leadership and discipline, two factors that can influence a unit's ability to accomplish its mission.

5. Let's start by discussing leadership. What aspects or types of leadership are important for a unit to accomplish its mission?
   a. *Probe:* How does an inspection team measure or assess them during an inspection, if at all?

6. What is good unit discipline?
   a. *Prompt as needed:* What does a unit with good discipline look like?
   b. How does an inspection team measure or assess unit discipline during an inspection, if at all?

7. How important is it to measure or assess unit leadership and discipline within the context of an inspection?
   a. How would you teach someone new to inspections, like me, to do this?
   b. How could the assessment of a unit's leadership and discipline be improved?

Let's shift gears a bit and discuss what happens during an inspection more broadly.

8. During an inspection, how long does it take for an inspection team to get a sense for how well a unit is doing? Generally speaking, is it a matter of minutes, hours, or days?
   a. *Probe:* How does this happen? What does the team consider?

9. What sources of information does an inspection team typically rely on during an inspection?
   a. Which do you think are the most useful and why?
   b. *Probe:* What about a unit's self-inspections? How helpful are they?

10. When it is time to write the report, how does an inspection team determine the overall grade for a functional area?
    a. *Prompt as needed:* What evidence or information serves as the basis for the findings a team reports?

We just have a few more questions. As I mentioned earlier, the Inspector General is considering a number of changes to the Air Force inspection system.

11. One type of change relates to the consolidation of inspections, assessments, and evaluations so that units have more inspection-free time during the year to focus on other aspects of their mission. This means, for example, that similar inspections required by different authorities would be conducted simultaneously, and redundant inspections would be reduced or eliminated. Why is this a good idea, or not?

12. How might such consolidation of many different inspections into a smaller number of more comprehensive ones affect an inspection team's work?
    a. *Probe:* What challenges might arise?
    b. *Probe:* How could such consolidation affect the interaction between a MAJCOM's IG and functional staff?

13. Another type of change under consideration involves relying more on a unit's self-inspections and less on external inspections, assessments, and evaluations. Why is this a good idea, or not?
    a. *Probe:* How much does the quality of unit self-inspection programs vary across units?
    b. *If comments indicate self-inspections are lacking:* How could self-inspection programs be improved?

14. A third change is IT-related: unit commanders would be provided with a standard inspection tracking and analysis toolset that gives them an automated way to monitor

unit compliance and track correction of deficiencies. The toolset would enable them to identify process deficiencies, assign actions to specific people, and conduct trend analysis. MAJCOM IGs and functional staffs would have visibility into the toolset and its updates. If wing commanders used such a toolset, would IG personnel regard this as an information source for the inspection process? Why or why not?

15. In closing, what would you like Air Force leadership to know about the compliance inspection process?
   a. What changes, if any, would you recommend?

Post-Inspection Leadership Element Interview Protocol

1. I know something about your background from the bios we received prior to this trip, but could you each tell me a little about your experience on inspection, assessment, and/ or evaluations teams? I'm interested in your experience as either IG core personnel or functional augmentees.

2. How much have the three [two] of you worked together, both in prior assignments and in this wing [group]?
   a. *If prior work together:* Please tell me more about the capacities in which you worked together before.

Let's start by talking about your recent UCI.

3. What did you and your airmen need to do to prepare for the UCI?
   a. How did the preparation affect your wing's [group's] day-to-day operations, if at all?
   b. To what extent did you use checklists based on AFIs or other instructions to prepare?
   c. What other data sources did you use? Examples include results from SAVs [staff assistance visits], past inspections, and your wing's self-inspection program.

4. Overall, what did you learn from the UCI?
   a. *Prompt:* What findings came as a surprise to you, if any?
   b. How did the IG's findings differ from the results of other inspections, evaluations, or assessments?
   c. How did the results compare to those from your own Self-Inspection Program?

5. How did—or will—this UCI change what your wing [group] is doing?
   a. *If changes noted:* Are the changes in response to what you learned during preparation, in response to areas for improvement noted by the IG team, or both?

6. What do think was more useful, preparation for the UCI, or the actual UCI itself, to include the IG's report, and why?

7. Let's talk more about your wing's [group's] self-inspection practices. We've learned that Self-Inspection Programs vary greatly not from unit to unit but also depending on

whether an inspection is coming up on the calendar. How much emphasis has it gotten in your wing [group], and how does that vary?

  a. ***Prompt:*** In your view, how extensive is it? Does that vary depending on whether an inspection is coming up on the calendar?

8. ***If Self-Inspection Program exists/is relatively extensive:*** What guidance or tools support it?

  a. To what extent do information systems play a role in your wing's Self-Inspection Program? By information systems, I mean any IT-based or automated tools that help with tasks like documenting deficiencies or tracking how and when they are addressed.

  b. How are self-inspection-related findings or actions documented?

  c. What types of self-inspection-related analyses are conducted in your wing? This could include analyses of patterns or trends over time.

  d. What self-inspection practices will continue now that the compliance inspection is over?

9. ***If Self-Inspection Program does not exist/is not used:*** How might a Self-Inspection Program help your wing?

10. To what extent do you evaluate the level of discipline within your wing [group], either in the context of a Self-Inspection Program or more generally?

  a. How do you define good unit discipline?

  b. How do you gauge or measure it?

11. To what extent do you evaluate leadership within your wing [group]?

  a. What leadership aspects or traits are important to ensure your wing [group] has good discipline and a high rate of compliance?

  b. How do you gauge or measure them?

12. Thinking not only in the context of inspections but more broadly, during regular operations, how do you learn about a failure to comply with an AFI or other instruction?

  a. ***Prompt:*** This could include, for example, mission briefs, standups, error reports, or airmen reports.

  a. What happens in such a situation? Please give me an example.

13. How important is it for your officers and enlisted personnel to be comfortable in reporting errors, instances of non-compliance, and other deficiencies that they observe or commit themselves?

14. ***If question twelve indicated importance:*** How comfortable are they?

  a. How do you know?

15. What actions can commanders take to create an environment in which people are comfortable coming to them to report mistakes, instances of non-compliance, and other deficiencies that could affect the mission?

I have one last set of questions. As I mentioned earlier, the Inspector General is considering a number of changes to the Air Force inspection system.

16. One type of change relates to the consolidation of inspections, assessments, and evaluations so that units have more inspection-free time during the year to focus on their mission. This means, for example, that similar inspections required by different authorities would be conducted simultaneously, and redundant inspections would be reduced or eliminated. Why is this a good idea, or not?
    a. What challenges might arise?

17. Another type of change under consideration involves relying more on a unit's self-inspections and less on external inspections, assessments, and evaluations. Why is this a good idea, or not?

18. If there were fewer external inspections in a calendar year, what effect might that have?
    a. *Prompt:* How might this influence unit effectiveness, either positively or negatively?
    b. What might a unit lose if it undergoes fewer external inspections?
    c. How often do you think a unit should have a UCI? Why?

19. If this shift occurred, what changes, if any, would be needed to your wing's inspection practices, including your Self-Inspection Program?

20. A third change is IT-related: unit commanders would be provided with MICT, a new, standard inspection tracking and analysis toolset that gives them an automated way to monitor unit compliance. The toolset would enable them to identify process deficiencies, assign actions to specific people, and conduct trend analysis. Why is this a good idea, or not?

21. What challenges might impede either its implementation or ongoing use?
    a. What obstacles, if any, do you foresee in integrating MICT with systems commanders already use?
    b. If the IG or MAJCOM leadership had visibility into such a toolset, how might this affect its use by commanders, if at all?

22. What should the Air Force do to encourage commanders to use a new tool like MICT?

23. In closing, what would you like Air Force leadership to know about the UCI process?
    a. What changes, if any, would you recommend?

Post-Inspection Wing Personnel Focus Group Protocol

1. We're going to begin by going around the room. Please each take a turn and tell us four things: your current pay grade, how long you've been in the Air Force, how long you've been assigned to this wing, and your role or primary responsibilities in it.

   Thanks for providing that information. Now that I know something about your individual backgrounds, I'll move into the questions for group discussion.

2. First, what is your overall reaction to the compliance inspection your wing recently underwent? Why do you feel that way?
    a. *Prompt:* We're interested in high level or overarching observations as opposed to a detailed list of pros and cons.

3. What did you need to do to prepare for the compliance inspection?
   a. How did the preparation affect your day-to-day operations, if at all?
   b. To what extent did you use checklists based on AFIs or other regulations to prepare?
   c. What other data sources did you use? Examples include results from SAVs, past inspections, and your wing's self-inspection program.
   d. What did your unit do in addition to preparation efforts led at the wing level?

4. Overall, what did you learn from the compliance inspection?
   a. *Prompt:* What findings came as a surprise to you, if any?
   b. How did the IG's findings differ from the results of other inspections, evaluations, or assessments?
   c. How did the results compare to those from your Self-Inspection Program?

5. How did the leadership in your wing affect its performance on the compliance inspection?
   a. *If leadership has some effect:* What leadership qualities or practices made a difference?
   b. *If leadership qualities identified:* How could you measure or assess them?

6. Let's talk more about your wing's self-inspection practices. We've learned that Self-Inspection Programs vary greatly not from unit to unit but also depending on whether an inspection is coming up on the calendar. How much emphasis has it gotten in your wing, and how does that vary?
   a. *Prompt:* In your view, how extensive is it? Does that vary depending on whether an inspection is coming up on the calendar?
   b. *If Self-Inspection Program exists/is relatively extensive:* How useful is the program? For example, has it revealed critical or significant deficiencies?
   c. *If Self-Inspection Program exists/is relatively extensive:* What self-inspection practices will continue now that the compliance inspection is over?
   d. *If Self-Inspection Program does not exist/is not used:* How might a Self-Inspection Program help your wing, or the functional area you work in in particular?

7. Thinking not only in the context of inspections, but more broadly, in your unit how willing and how open are people to reporting mistakes, deficiencies, or compliance-related problems?
   a. Why do you think it's like that in your unit?
   b. In your view, what do you think is needed to create an environment in which Airmen feel comfortable coming to leadership with concerns like these?

We just have a few more questions. As I mentioned earlier, the Inspector General is considering a number of changes to the Air Force inspection system.

8. One type of change relates to the consolidation of inspections, assessments, and evaluations so that units have more inspection-free time during the year to focus on their mission. This means, for example, that similar inspections required by different authorities

would be conducted simultaneously, and redundant inspections would be reduced or eliminated. Why is this a good idea, or not? What are the pros and cons?

   a. What challenges might arise?

9. Another type of change under consideration involves relying more on a unit's self-inspections and less on external inspections, assessments, and evaluations. Why is this a good idea, or not? What are the pros and cons?

   a. What changes, if any, would be needed to the Self-Inspection Program if this shift occurred?

   b. What might a unit lose if it undergoes fewer external inspections?

   c. How often do you think a unit should have a UCI? Why?

10. A third change is IT related: unit commanders would be provided with a new, standard inspection tracking and analysis toolset that gives them an automated way to monitor unit compliance. The toolset would enable them to identify process deficiencies, assign actions to specific people, and conduct trend analysis. Why is this a good idea, or not?

   a. ***If viewed as a good idea:*** What should the Air Force do to encourage commanders to use this new tool?

   b. If the IG or MAJCOM leadership had visibility into such a toolset, how might this affect its use by commanders, if at all?

11. In closing, what would you like Air Force leadership to know about the compliance inspection process?

   a. What changes, if any, would you recommend?

Risk Management in the Federal Aviation Administration (FAA) Inspection System

When FAA Principal Inspectors discover potential safety hazards (either as part of the current certification cycle or based on information received from various sources between the inspection cycles), they can initiate the Risk Management Process (RMP) to perform a more in-depth risk assessment for the specific hazard item(s). PIs can use the RMP to document, track, and manage hazards and their associated risks. The RMP can address any hazard that the PIs decide is significant enough to justify analysis and tracking. There is no mandatory requirement for its use.

The RMP has five major steps:

1. Identify the hazard.
2. Analyze and assess the risk associated with the hazard.
3. Make a decision about how to address the risk.
4. Implement the decision.
5. Validate the effectiveness of the decision.

FAA PIs can apply this process in the regular SAI or EPI certification processes described in Chapter Two. They can also use it to schedule additional inspections and allocate additional resources for those items that have potentially high risks. As described in the following sections, PIs can apply additional inspections and resource allocations in each of the five steps of the RMP process.

Step 1: Identify Hazards

A hazard is a condition, event, or circumstance that could lead or contribute to an unplanned or undesired event. In the RMP, the PIs first identify any hazard in the carrier's operating environment or systems. The PIs analyze data from many sources to determine if hazards are isolated incidents or systemic problems. They focus on systemic hazards and their potential consequences to determine the level of risk associated with any hazard. Without conducting a complete analysis, the PIs may notify a certificate holder of any isolated incidences that do not require a complete RMP. If the isolated incident leads to a determination of noncompliance, then national guidance must be followed for processing enforcement action.

Once PIs identify a systemic hazard, they prepare a summary that describes the hazard, including relevant facts such as what the hazard is, why it occurs, how often it occurs, where it occurs, and who is responsible for it. The PIs also determine and document the potential consequences that could result if the hazard is not addressed or corrected. These consequences could be any one of the following: (1) equipment failure, (2) human error, (3) damage to equipment, (4) procedural nonconformance, (5) process breakdown, (6) personal injury or death, (7) regulatory noncompliance, (8) decreased quality or efficiency, or (9) other.

Step 2: Analyze and Assess Risk

The risk analysis approach used in RMP mirrors the traditional PRA method of identifying (1) possible outcomes (risk factors associated with the identified hazards), (2) the probability of these outcomes (the likelihood value of risk factors), and (3) a value assessment of the consequences (the severity of the consequences).

After PIs have identified hazards, they conduct a risk analysis to analyze and identify risk factors associated with the hazards. Risk factors are typically situational factors (e.g., operating conditions that promote corrosion, aging aircraft, or high-cycle use of aircraft) or deficiencies in design or performance related to safety attributes (e.g., missing attributes or failure to adhere to procedures). Risk factors identify what must later be controlled and mitigated to reduce the overall level of risk. An effective action plan should address risk factors by eliminating them or by reducing their impact.

The PIs determine whether there are known risk factors associated with the severity of the consequences and the likelihood of their occurrence. When risk factors are unknown, the PIs suspend the RMP and conduct additional research (including additional data collection through inspections) on the risk factors before assessing the risk. The PIs may use SAI, EPI, or other means to obtain more information about the factors affecting the level of risk.

Once the PIs have identified and documented risk factors, they determine the appropriate value related to the severity of the potential consequences. Using a combination of available data and expert judgment, they determine whether the severity is:

1. High: The consequences threaten a potential loss (or breakdown) of an entire system or subsystem or an accident or incident.
2. Medium: The consequences threaten potentially moderate damage to an aircraft, partial breakdown of an air carrier system, or violation of regulations or company rules.
3. Low: The consequences threaten potential poor air carrier performance or disruption to the air carrier.

The PIs also determine the appropriate value related to the likelihood of the consequences actually occurring. PIs similarly assess likelihood using a combination of available data and expert judgment to determine whether consequences are:

1. frequent—continuously experienced
2. probable—occur often
3. occasional—occur several times
4. remote—unlikely, but could occur.

The PI considers the overall level of risk to determine the priority of ensuring that the carrier addresses the hazard and its associated level of risk. This assessment assists the

PIs in decisionmaking, action planning, and evaluating air carrier actions. The PIs use the likelihood-severity risk matrix in Figure C.1 to determine the overall level of risk associated with the identified hazards.

Step 3: Perform Decisionmaking

Based on the results of the risk assessment in Step 2, the PIs decide on one of the following actions: (1) eliminate the hazard, (2) mitigate the risk, (3) accept the risk at its existing level, or (4) transfer the risk. When corrective action is beyond the PIs' authority—e.g., actions such as rule changes, new or revised airworthiness directives, or policy changes—the PIs can transfer the authority, responsibility, and accountability for taking corrective action for the identified hazard to the appropriate FAA organization. Also, where the overall level of risk falls into the blue area of the risk matrix, PIs may accept it without further action.

  If the overall level of risk is found to be unacceptable, however, the PIs document the mitigation rationale based on the following: If the overall level of risk falls into the red area, the PIs assess the risk as unacceptable and initiate further work to eliminate the associated hazard or control the factors that lead to higher risk likelihood or severity. If the risk assessment falls into the yellow area, the PIs can accept the risk under defined conditions of mitigation. An example of this situation would be an assessment of the impact of an inoperative aircraft component that is deferred in accordance with a minimum equipment list (MEL). Defining an operational or maintenance procedure in the MEL would constitute a mitigating action that could make an otherwise unacceptable risk acceptable, as long as the defined procedure was implemented.

Figure C.1
Effects of Likelihood and Severity on Level of Risk

Risk Matrix			
Likelihood	Severity		
	High	Medium	Low
Frequent	1	3	5
Probable	2	6	8
Occasional	4	9	11
Remote	7	10	12

NOTES: 1–3 (red) = high overall risk; 4–9 (yellow) = medium overall risk; 10–12 (blue) = low overall risk.
RAND *TR1291-C.1*

Step 4: Implement the Decision

The PIs implement the mitigation strategies chosen in Step 3 to ensure that the carrier addresses the identified hazard and unacceptable levels of risk. With PI oversight, a carrier is usually responsible for executing the mitigation strategies. The PIs then identify the FAA actions and resources needed to oversee the carrier's implementation of the strategies. Sometimes, the PIs select strategies that do not involve the participation of the carrier (e.g., reevaluating carrier program approvals, authorizations, deviations and exemptions; amending or revoking the carrier's authority to conduct all or part of an operation; or initiating an enforcement action).

As a part of this step, the PIs also develop action items that address the risk factors. Action items describe how, where, and when an action should be done and may include (1) reevaluating the carrier's programs, approvals, authorizations, deviations, and exemptions; (2) amending or revoking the carrier's authority to conduct all or part of its operation; (3) initiating an enforcement investigation; (4) suspending the certification process; or (5) convening a technical team for additional analysis. The PIs also assign personnel who can perform the action items and monitor the progress until the RMP is closed for a particular hazard. The PIs ensure that (1) all action items are complete, and that (2) the current data indicate that the action plan has eliminated the hazard or reduced the associated risk to an acceptable level.

Step 5: Validate the Effectiveness of the Decision and Close the RMP

After all action items are complete, or data indicate that the action plan has eliminated the hazard or reduced the associated risk to an acceptable level, the PIs validate the effectiveness of the selected approach. The PIs review the status of the hazard to verify that the carrier has eliminated the hazard or mitigated the level of risk associated with the hazard to an acceptable level. After evaluating the results of the mitigation strategies, the PIs decide whether to close the RMP or to require the development and implementation of additional action items. These additional action items may trigger additional inspections and resource allocations. After determining that all risk factors have been addressed to the extent possible, the PIs also review the hazard and its consequences to revise the severity and likelihood values and overall risk ratings as necessary. When it is determined that the risk level is acceptable, the PIs close the RMP and monitor the relevant hazard through DA and PA as part of the current or future certification cycles.

Additional Background on the Air Force Climate Survey

This appendix provides more detailed information on the 2003 Air Force Climate Survey. The opening section on "Duty Information" lists questions and available answers. All other sections list only questions. Unless indicated otherwise below, these sections use appropriate Likert scales to structure answers to the questions in them. The text below is extracted from the original survey verbatim. The survey imbeds the text shown here in fairly extensive background material and instructions.

Duty Information

Unit Number <u>(0-9)</u> <u>(0-9)</u> <u>(0-9)</u> <u>(0-9)</u> <u>(0-9)</u>
Break Number <u>(0-9)</u> <u>(0-9)</u> <u>(0-9)</u> <u>(0-9)</u> <u>(0-9)</u>
English ☐   *Non-English* ☐

1. Select your **primary duty** with the Air Force. Your primary duty is defined as the capacity in which you spend 51 percent or more of your time performing your duties. Air Reserve Technicians (ARTs) and Technicians (TECHs) should select Air Force Reserve or Air National Guard as appropriate for your primary duty, not Appropriated Fund civilian (Civil Service).
   a. Active Duty (go to 4)
   b. Air Force Reserve (go to 2)
   c. Air National Guard (go to 3)
   d. Air Force Appropriated Fund civilian (Civil Service) (go to 5)
   e. Air Force Non Appropriated Fund (NAF) Civilian (e.g., Services Employees) (go to 5)

2. If you are in the Air Force Reserve, select the **classification** that best describes your role.
   a. Traditional
   b. Active Guard/Reserve (AGR)/Statutory (Stat) Tour (go to 4)
   c. Air Reserve Technician (ART) (go to 4)
   d. Individual Mobilization Augmentee (IMA) (go to 4)

3. If you are in the Air National Guard, select the **classification** that best describes your role.
   a. Traditional
   b. Active Guard/Reserve (AGR)/Statutory (Stat) Tour
   a. Technician

4. Please select the military **category** that best describes your role.
   a. Officer
   b. Enlisted

5. Select the item that best describes your present **duty status**. This should best describe where you currently work on day-to-day basis.
   a. At my home station (including matrixed personnel) and not in student status (start the Climate Questions)
   b. TDY and not in student status (start the Climate Questions)
   c. Student Status (PME or Commercial) (go to 6)
   d. Deployed, Mobilized, or Activated (go to 7)

6. While attending this training, I am
   a. At home station or on TDY orders to attend training (start the Climate Questions)
   a. PCS'd [permanent change of station] to attend training (see paragraph below)

7. Please select the Area of Responsibility (AOR) in which you are currently assigned.
   a. Central Command (CENTCOM)/Central Air Forces (CENTAF)
   b. European Command (EUCOM)/United States Air Forces in Europe (USAFE)
   c. Joint Forces Command (JFCOM)
   d. Northern Command (NORTHCOM)/Northern Air Forces (NORTHAF)
   e. Pacific Command (PACOM)/Pacific Air Forces (PACAF)
   f. Southern Command (SOUTHCOM)/Southern Air Forces (SOUTHAF)
   g. Don't know/can't say

Job

1. My job requires me to use a variety of skills.
2. My job allows me to see the finished products of my work.
3. Doing my job well affects others in some important way.
4. My job is designed so that I know when I have performed well.
5. My job allows me freedom to work with minimum supervision.

Resources

1. I have adequate time to do my job well.
2. We have enough people in my work group to accomplish the job.
3. I have the right tools/equipment to accomplish my job.
4. I have enough time to accomplish my daily workload during my duty hours.

Core Values

1. I am able to do my job without compromising my integrity.
2. Overall, people in my unit uphold high standards of excellence.

3. Overall, people in my unit demonstrate that duty takes precedence over personal desires.
4. Overall, people in my unit are held accountable for behavior that contradicts the AF core values.

Supervision

1. My supervisor is good at planning my work.
2. My supervisor sets high performance standards.
3. My supervisor is concerned with my development.
4. My supervisor corrects poor performers in my work group.
5. My supervisor looks out for the best interests of my work group.
6. My supervisor provides instructions that help me meet his/her expectations.
7. My supervisor helps me understand how my job contributes to my unit's mission.
8. My supervisor ensures that there is a fair distribution of the workload among the people.
9. My supervisor provides opportunities for me to give feedback to him/her.

Unit Leadership

1. The leaders in my chain of command (in my unit) listen to my ideas.
2. The leaders in my chain of command (in my unit) are easily accessible.
3. I trust the leaders in my chain of command (in my unit).
4. I am proud to be associated with the leaders in my chain of command (in my unit).
5. I see the leaders in my chain of command (in my unit) doing the same things they publicly promote (walking the talk or leading by example).

Leadership Behaviors

1. My unit commander (or commander equivalent) sets challenging unit goals.
2. My unit commander (or commander equivalent) provides a clear unit vision.
3. My unit commander (or commander equivalent) makes us proud to be associated with him/her.
4. My unit commander (or commander equivalent) is consistent in his/her words and actions.
5. My unit commander (or commander equivalent) is inspirational (promotes esprit de corps).
6. My unit commander (or commander equivalent) motivates us to achieve our goals.
7. My unit commander (or commander equivalent) is passionate about our mission.
8. My unit commander (or commander equivalent) challenges us to solve problems on our own.
9. My unit commander (or commander equivalent) encourages us to find new ways of doing business.
10. My unit commander (or commander equivalent) asks us to think through problems before we act.
11. My unit commander (or commander equivalent) encourages us to find innovative approaches to problems.

12. My unit commander (or commander equivalent) listens to our ideas.
13. My unit commander (or commander equivalent) treats us with respect.
14. My unit commander (or commander equivalent) is concerned about our personal welfare.

Unit Commander Behavior Feedback

1. Integrity: Consistently adhering to a moral or ethical code or standard. A person who considers the "right thing" when faced with alternate choices.
2. Organizational Loyalty: Being devoted and committed to one's organization.
3. Employee Loyalty: Being devoted and committed to one's co-workers and subordinates.
4. Selflessness: Being genuinely concerned about the welfare of others and willing to sacrifice one's personal interest for others and our organization.
5. Compassion: Concern for the suffering or welfare of others and provides aid, or shows mercy for others.
6. Competency: Capable of executing responsibilities assigned in a superior fashion and excels in all task assignments. Is effective and efficient.
7. Respectfulness: Shows esteem, consideration, and appreciation of other people.
8. Fairness: Treats people in an equitable, impartial, and just manner.
9. Self-Discipline: Can be depended upon to make rational and logical decisions (in the interest of the unit).
10. Cooperativeness: Willingness to work or act together with others in accomplishing a task or some common end or purpose.
11. Sociability: Acts in an enthusiastic, friendly, and courteous manner toward others. Communicates in tactful and diplomatic ways. Provides a positive atmosphere.

Organization Characteristics

1. If you were released from all of your service obligations and you could separate from the Air Force within the year, what is the likelihood that you would leave the Air Force?
2. I find that my values and the organization's values are very similar.
3. I am proud to tell others that I am part of this organization.
4. There's not too much to be gained by sticking with this organization until retirement (assuming I could do so if I wanted to).
5. Often, I find it difficult to agree with the policies of this organization on important matters relating to its people.
6. Becoming a part of this organization was definitely not in my best interest.
7. How many hours do you perform actual work for the Air Force in a typical week at your home station? (Fill in the blank: (0–1) (0–9) (0–9) hours per week)
8. Answered only if you are deployed: How many hours did you perform actual work for the Air Force in a typical week at your deployed location? (Fill in the blank: (0–1) (0–9) (0–9) hours per week)
9. At your current duty location, do you work "overtime," that is, longer than the stated normal duty hours? (Yes No [skip to Training and Development])

10. How many hours do you work "overtime," that is, over the stated normal duty hours? (Fill in the blank: (0–1) (0–9) hours per week)

Training and Development

1. I am given opportunities to improve my skills.
2. I am encouraged by my unit leadership to learn new things.
3. I have been adequately trained for the job I am expected to do.
4. I am allowed to attend continuing professional training (workshops, conferences, etc.).

Teamwork

1. People in my work group respect each other.
2. My work group adequately resolves conflicts.
3. Members of my work group willingly share information.
4. People in my work group cooperate to get work done.

Participation/Involvement

1. I feel free to suggest new and better ways of doing things.
2. I am asked how we can improve the way my work group operates.
3. Sufficient effort is made to get the opinions and ideas of people in this work unit.
4. Suggestions made by unit personnel are implemented in our daily work activities.

Recognition

1. My unit's leaders reward team performance fairly.
2. My unit's leaders reward individual performance fairly.
3. When deserved, my unit's leaders do a good job of recognizing people in all grades and types of jobs.
4. My unit's leaders reward primary job expertise more than additional duty performance.

Unit Flexibility

1. My unit adapts to changes quickly.
2. My unit encourages appropriate risk taking.
3. My unit challenges old ways of doing business.
4. My unit adapts to changes well.

General Satisfaction

1. In general, I am satisfied with my job.
2. I have a sense of personal fulfillment at the end of the day.
3. The tasks I perform provide me with a sense of accomplishment.
4. I am a valued member of my unit.

5. I would recommend an assignment in my unit to a friend.
6. Morale is high in my unit.

Unit Performance Outcomes

1. The quality of work in my unit is high.
2. The quantity of work accomplished in my unit is high.
3. My unit is known as one that gets the job done well.
4. My unit is successfully accomplishing its mission.

Job Enhancement

1. In my unit, people help each other out when they have heavy workloads.
2. In my unit, people make innovative suggestions for improvement.
3. In my unit, people willingly give of their time to help members who have work-related problems.
4. In my unit, people willingly share their expertise with each other.

Historical

1. I was in this unit when the 2002 Air Force Chief of Staff Climate Survey results were released in May 2002. (Yes No)
2. My unit leader(s) used the 2002 Air Force Chief of Staff Climate Survey results in a positive way.

Additional Background on the Air Force Culture Assessment Safety Tool (AFCAST)[1]

This appendix provides background on the ideas about human errors and high reliability organizations that underlie the questions in AFCAST surveys. It then provides a list of the questions in the current versions of the four non-nuclear surveys.

The Theoretical Basis for AFCAST Survey Design

AFCAST surveys seek to assess the organizational factors that affect safety and performance. To do this, AFCAST draws on a literature that emphasizes the role of human error in safety and performance mishaps and attributes of complex organizations that affect the incidence of human errors. Human error has been shown to be the single largest contributing factor in industrial accidents and failures. For example, human error has been implicated in 70 to 80 percent of all civil and military aviation accidents (Shappell and Wiegmann, 1996). Empirical research has studied five types of factors that might explain such errors:[2]

- Cognition: Errors occur when individuals in a complex system fail to process information passing among them correctly.[3]
- Ergonomics and system design: Errors occur when humans do not interact appropriately with hardware, software, and inputs from their environment in a complex system (Edwards, 1988).
- Aeromedical: Errors result from the physiological status of a pilot (e.g., effects of fatigue, illness, and medications).[4]
- Psychosocial: Errors occur when group dynamics degrade interpersonal communication.[5]
- Organizational: Errors result from poor organizational designs, policies, or practices.

---

[1] This appendix draws on the references cited and on discussions with individuals responsible for designing and managing the surveys described here.

[2] For a review of the literature, see Wiegmann, Rich, and Shappell, 2000. See also Shappell and Wiegmann, 1996; Wiegmann and Shappell, 1997; Shappell and Wiegmann, 2000; Wiegmann and Shappell, 2000; Shappell and Wiegmann, 2001; Wiegmann and Shappell, 2001.

[3] Rasmussen, 1982; O'Hare et al., 1994.

[4] National Transportation Safety Board, 1994; Reinhart, 1996.

[5] Helmreich and Foushee, 1993; Federal Aviation Administration, 1997; Wiegmann and Shappell, 1999.

AFCAST surveys grew from efforts to understand this last factor. The most mature model available to understand how organizational problems contribute to human error is the "Swiss cheese" model documented by Reason (1990).[6] Reason argued that errors occur when four different kinds of failures in an organization align—like holes in slices of Swiss cheese stacked on top of one another—so that a set of latent conditions allow an active human behavior or decision to result in an accident. The accident occurs only if the following failures occur in all the right places in all four layers of defense at the same time:

- organizational factors (fallible decisions of high-level decisionmakers yielding, e.g., poor training, policies, and leadership development)
- unsafe supervision (line management deficiencies, e.g., poor leadership, command and control)
- preconditions for unsafe acts (psychological precursors to unsafe acts, e.g., fatigue, conflicting personalities, poor situational awareness, poor teamwork)
- unsafe acts (e.g., failure to follow standard procedure, instructions).

Note that, even though efforts to address safety issues provided the basis for this model, there is nothing about it that limits its application to safety per se. These latent and active factors are relevant wherever the problems highlighted in the model can align in a complex organization (like an Air Force wing).

Shappell and Wiegmann (2000) built on Reason's work to provide a detailed taxonomy of sources of human error (i.e., the holes in the cheese) at each level for use in accident investigation and mishap analysis. It is called the Human Factors Analysis and Classification System. Today, DoD uses a version of HFACS (DoD HFACS) to assess safety and accidents.[7] The DoD Human Factors Working Group of the Joint Services Safety Chiefs reviews and updates DoD every six months.

The HFACS framework has been applied, for example, to the military aviation, commercial aviation (air carrier, commuter, and general aviation), and mining industries to analyze accident reports and develop accident prevention strategies. It is interesting to note that, in most of these studies, organizational influences have not been readily apparent. This work has revealed that organizational influences and unsafe supervisions appeared to have had relatively insignificant impact compared to the lower tier causes such as preconditions for unsafe acts and unsafe acts.[8]

When the Air Force Safety Center initially fielded AFCAST in 2007, AFCAST tools and survey design were based largely on survey questionnaires used in the Model of Organizational Safety Effectiveness (MOSE). MOSE derived primarily from the theory of High Reliability Organizations (HROs) pioneered by Karlene H. Roberts,[9] whose work owed a great deal to the same body of work underlying Reason's study (1990). HRO theory seeks to identify the key attributes of organizations that operate in a hazardous environment, but have very low rates of accidents and other adverse incidents. Roberts (1990) used experience in air traffic control,

---

[6]   Reason, 1990; Reason, 1997. Related work includes Weaver, 1971; Adam, 1976; Bird and Loftus, 1976; Heinrich, Peterson, and Roos, 1980; Shappell and Wiegmann, 2000.

[7]   Department of Defense Human Factors Analysis and Classification System, undated.

[8]   Wiegmann and Shappell, 2001; Shappell and Wiegmann, 2001; Patterson and Shappell, 2010.

[9]   Roberts, 1990; Roberts, 1993; Libuser, 1994; Roberts, Rousseau, and La Porte, 1994.

nuclear power plants, and U.S. Navy aircraft carriers to develop the theory. Roberts found that these organizations have certain key characteristics in common, including sound safety management policies, standardized procedures, adequate resources and staffing, defined systems of risk management, and strong leadership styles.

Cultural factors are difficult to define in terms amenable to observation and measurement, but researchers at that Naval Postgraduate School in Monterey, California, developed MOSE to incorporate the aspects of organizational climate that underlie naval aviation values and norms.[10] MOSE has since been modified and expanded for use in other industries, including commercial aviation, healthcare, and finance.[11]

Anthony Ciavarelli and his colleagues developed a web-based system of online survey questionnaires designed to assess the possible influences of organizational factors such as organizational climate, safety culture, workload, and resource availability on individual behavior in naval aviation. They used the five major areas identified in the HRO theory to formulate the survey questionnaires:

1. *process auditing*—the organizational system of checks that identifies hazards and the means to correct safety problems when they are identified
2. *culture and reward system*—the expected social rewards and disciplinary actions used to reinforce safe behavior and correct unsafe behavior
3. *quality assurance*—the policies and procedures for promoting a high quality of work performance
4. *risk management*—comprising accurate risk perception and a systematic process to identify hazards and control operational risks
5. *leadership and supervision*—leaders openly committed to safety who actively promote a strong safety culture ensure that resources, policies, plans, processes, and the selection and training of personnel maintain safe and successful operations.

The MOSE surveys use a Likert scale to structure answers to a set of questions chosen to measure respondent perceptions about factors relevant to each of these five areas.[12] The designers of these surveys considered protection of the respondents and participating organizations to be critical to obtaining candid inputs, so the surveys protected the identity of respondents and their responses.

Although the principal areas addressed by an HRO grew from looking at safety issues in organizations facing high-risk environments, there is nothing about HRO theory that restricts its application to traditional safety issues. The concerns identified in all five areas are relevant wherever hazards generate significant risks that a complex organization must manage reliably to succeed. Applications of the MOSE model to organizations with risk concerns other than safety, like banks, have confirmed that the basic approach has broad applicability. The MOSE model can be particularly helpful in assessing the effects of leadership and organizational discipline on organizational performance when risk management is important to the organization's effectiveness. But it must be matched to the context in which it will be applied. The empirical

---

[10] Ciavarelli and Figlock, 1997; Ciavarelli et al., 2001; Ciavarelli and Crowson, 2004.

[11] Ciavarelli, 2005; Ciavarelli, 2007b.

[12] Options included "strongly disagree," "disagree," "neutral," "agree," "moderately agree," "strongly agree," and "not applicable."

match is more important to success than ensuring any interpretation of a "pure" application of the underlying principles.

Over time, MOSE survey results have revealed empirical findings that indicate that (1) higher-level personnel in an organization consistently tend to perceive a higher quality safety climate than lower-level personnel; (2) quality levels of organizational safety climates differ significantly across industries (e.g., healthcare respondents have consistently rated the quality of their organizational safety climate significantly lower than did respondents in selected U.S. Navy activities); and (3) for some activities, organizations with a higher quality organizational climate tend to perform better than analogous organizations with a lower quality organizational climate.[13]

These findings illustrate the uncertainties related to interpreting surveys of perceptions about organizational safety: Do lower-level personnel rate safety lower than their leaders because they have a better knowledge of the actual level of safety in an organization or because higher-level personnel have a better understanding of organizational safety goals and where they fit in the broader setting of organizational goals? Is the level of safety in health care activities really lower than that in the Navy activities observed, or do personnel in a health care setting judge the level of safety that they perceive against higher standards or expectations than personnel involved in the Navy activities? And why is the relationship between perceived safety and broader organizational performance so much stronger in some settings than in others? The findings of HRO-based surveys cannot address such questions by themselves. Users must view these findings in a broader context to understand useful policy implications of surveys results. In particular, such surveys cannot replace more traditional safety inspections, but they can help target hands-on inspections to help assess anomalies detected by a survey.

AFSC based the structure and content of its first AFCAST surveys on the Aviation Command Safety Assessment Survey Questionnaire developed by MOSE for the U.S. Navy and Marine Corps. Since 2007, AFSC has repeatedly modified the AFCAST surveys to make them more responsive to the priorities of potential users. Experience has taught AFSC that it is helpful to align the surveys with the framework that the Air Force uses elsewhere to assess safety—the HFACS framework described above. Doing so makes it easier for squadrons to use AFCAST findings to do diagnostics, which, given DoD guidance, must draw on HFACS-based databases. It also facilitates communication between the AFCAST program and the Aviation Safety Analysis System (ASAS) that the Federal Aviation Administration uses to collect, process, and disseminate safety-related information. Experience with the MOSE approach and input from potential users have led AFSC to focus current AFCAST surveys on the first two parts of HFACS: organizational influences and unsafe supervision.

The questionnaires in the current AFCAST surveys are organized into four broad categories: organizational processes, organizational climate, resources, and supervision. The first three are components of the organizational influences in HFACS. The last captures the effects of unsafe supervision. Individual questions in AFCAST, however, still cover various aspects of the five major factors emphasized in HRO theory. As AFSC continues to refine the AFCAST surveys, it mixes and matches elements of the MOSE and HFACS approaches to combine the conceptual insights of this still active body of analysis with the policy priorities of the Air Force leaders who use the surveys.

---

[13] Gaba et al., 2003; Ciavarelli and Crowson, 2004; Desai, Roberts, and Ciavarelli, 2006; Ciavarelli, 2007a; Ciavarelli, 2007b.

Current Design of AFCAST

Table E.1 displays the questions in the current version of the non-nuclear AFCAST surveys. The columns on the right show the number of each question in each of the operations (Ops), maintenance (Mx), support (Sup), and higher headquarters (HHQ) surveys. If a cell in any of these columns has a number in it, a close analog of the question in the same row appears in the survey.

As noted in Chapter Five, the IG might ask very similar questions in its own version of such a survey but with a different emphasis in the first three categories in the table—again, based on HFACS organizational influences—as well as in the fourth category highlighted in Chapter Five and based on the consideration in HFACS unsafe supervision. Rather than focusing on safety, the IG could highlight readiness, compliance, effectiveness, leadership, and discipline in a unit. Questions shown in blue would require no change at all. Those shown in green could apply to a broader IG scope with only minor changes. The questions in orange would require more substantial changes to relate to the IG's mission. Only the questions in red would be hard for the IG to adapt. To understand the ease of reframing such questions, remember that a hazard, adverse incident, or human error need not have anything to do with physical safety. Risk management and quality assurance can be effectively applied to control a broad range of potential failures that have nothing to do with safety per se.

Table E.1
AFCAST Survey Questions for Operations, Maintenance, Support, and Higher Headquarters

	Question, to Be Answered in a Likert Scale	Ops	Mx	Sup	HHQ
Organizational Processes	My squadron adequately reviews and updates safety standards and operating procedures.	1	1	1	
	My squadron closely monitors job qualifications [currency standards].	2	2	2	
	My squadron adequately trains our [my directorate/division's] personnel to safely conduct their jobs.	3	3	3	1
	My squadron recognizes individual safety acts through awards and incentives.	4	4	4	2
	My squadron routinely meets or exceeds its operational training goals.	5			
	Safety decisions are made at the proper levels by the most qualified personnel.	6	5	5	3
	Standards in my squadron [directorate/division] are clearly defined.	7	6	6	4
	Standards in my squadron [directorate/division] are enforced.	8	7	7	5
	My squadron makes effective use of the flight surgeon to help identify and manage high risk personnel.	9			
	My squadron temporarily restricts operators from flying/pulling missile alerts/conducting space missions who are under high personal stress.	10			
	Operators in my squadron are given qualifications [increased responsibility] without the appropriate experience or skills.	11	8	8	6
	Anyone intentionally violating standard operating procedures (SOPs) or safety rules is swiftly corrected.	12	9	9	7
	My squadron's operating standards when deployed are of the same quality as our operating standards when at home base	13		10	
	Work performance when deployed is of the same quality as our work performance when at home base.		10		

Table E.1—Continued

	Question, to Be Answered in a Likert Scale	Ops	Mx	Sup	HHQ
	Official guidance (e.g., AFIs, TOs) is incorporated into day-to-day safety decisions in my squadron.	14	11	11	
	My squadron accurately identifies and assesses hazards associated with its flight/missile/space operations.	15	12	12	
	My squadron adequately monitors daily operations to catch possible human errors.	16	13	13	
	Squadron members [individuals], from the top down, incorporate risk management into daily activities.	17	14	14	8
	My squadron's Crew Resource Management (CRM) program is helping to improve mission performance and safety.	18			
	Effective communication flow exists within my squadron.	19	15	15	9
	Effective communication flow exists with external squadrons.	20	16	16	10
	My squadron effectively communicates pertinent information during shift changes.		17	17	
	Tool control is closely monitored.		18		
	Maintenance records are accurately maintained in my squadron.		19		
	Work in my squadron is supervised and staffed by qualified personnel.		20	18	
	Workers are briefed on potential hazards associated with their assigned tasks in my squadron.		21	19	
	Our Safety directorate/division keeps me well informed regarding relevant hazards/mishaps.				11
	My directorate/division provides adequate oversight of similar directorate/divisions in subordinate commands.				12
	My directorate/division provides adequate assistance to similar directorate/divisions in subordinate commands.				13
	In my squadron, Stan/Eval and check rides are conducted as intended, to honestly assess aircrew/missile and space crews' qualifications.	21			
	Aircrew/missile crews/space crews in my squadron are encouraged to submit and discuss Aircraft discrepancies with Maintenance Operations Control (MOC)/ICMB discrepancies with MOC/space anomalies with senior leadership and/or higher headquarter before and after flights/missile alert tours/space missions (pre-launch, launch and orbital mission areas).	22			
Organizational Climate	My squadron [headquarters] has a reputation for high-quality performance.	23	22	20	14
	Violations of operating procedures, flying/missile/space regulations, or general flight discipline [AFIs/TOs/procedures and regulations] are rare [in my squadron].	24	23	21	15
	Our squadron [headquarters] conceals adverse incidents.	25	24	22	16
	Training is often postponed/cancelled.	26	25	23	17
	Individuals are comfortable approaching supervisors about personal problems/illness.	27	26	24	18
	Individuals in my squadron are willing to report safety violations, unsafe behaviors, or hazardous conditions.	28	27	25	19
	Quality Assurance (QA)/Quality Assurance Evaluator (QAE) positions are desirable assignments in my squadron.		28		
	[QA/QAE/The Safety directorate] Stan/Eval is a well-respected element of my squadron.	29	29		20

Table E.1—Continued

	Question, to Be Answered in a Likert Scale	Ops	Mx	Sup	HHQ
	Safety days are effective in my squadron.	30	30	26	21
	Squadron members feel pressured to cut corners to accomplish their job/mission.	31	31	27	22
	Conflicts between members degrade performance within my squadron.	x	32	28	23
	[Duty shifts and] crew rest policies are enforced in my squadron.	33	33	29	
	Our personnel [Members of my directorate/division] work effectively as a team.	34	34	30	24
	The Flight Safety Officer (FSO)/Missile/Space Safety Officer position [NCO/Unit Safety Representative positions] is a desirable position in my squadron.	35	35	31	
	My directorate/division has a good working relationship with other directorate/divisions in my headquarters.				25
	My directorate/division has a good working relationship with similar directorate/divisions in subordinate organizations.				26
	Morale in my squadron is high.	36	36	32	27
Resources	I am provided adequate resources (e.g., time, staffing, budget, and equipment) to accomplish my job.	37	37	33	28
	Based upon our current manning/assets, my squadron [directorate/division] is over-committed.	38	38	34	29
	My squadron provides me with the right number of flight/missile/space training hours per month to operate safely.	39			
	I have adequate time to prepare for and brief my flights/missile alert tours/space missions.	40			
	Fatigue (due to operational demands) is degrading performance in my squadron.	41	39	35	30
	Fatigue (due to life style, behavior, and judgment) degrades performance in my squadron.	42	40	36	
	My squadron has sufficient experienced personnel [staffing] to operate safely.	43	41	37	31
	Night crew has sufficient staffing to meet workload demands in my squadron.		42		
	Required publications are current and used in my squadron.		43	38	32
	Required tools [computers] and equipment are serviceable and used in my squadron.		44	39	33
	Parts are sufficiently available to meet maintenance demands.		45		
	Additional duties adversely affect my performance in my squadron.	44	46	40	34
	Temporary duty (TDY) deployment rates for the last year created safety problems in my squadron.	45	47	41	35
Supervision	Leaders/supervisors in my squadron are actively engaged in the safety program and management of safety matters.	46	48	42	36
	Leaders/supervisors in my squadron balance safety concerns with achieving mission tasking.		49	43	
	Leaders/supervisors are more concerned with operational tasks than safety.				37
	Leaders/supervisors encourage reporting safety discrepancies without fear of negative repercussions.	47	50	44	38
	Leaders/supervisors in my squadron set a good example for compliance with policies, rules, and instructions.	48	51	45	39

Table E.1—Continued

	Question, to Be Answered in a Likert Scale	Ops	Mx	Sup	HHQ
	Leaders/supervisors in my squadron permit cutting corners to get a job done.	49	52	46	40
	Leaders/supervisors in my squadron react well to unexpected changes.	50	53	47	41
	Leaders/supervisors in my squadron care for members' quality of life.	51	54	48	42
	The Flight Safety Officer (FSO)/Missile/Space Safety Officer [squadron Safety Office/Safety NCOs/Safety Representatives/Safety personnel] is effective at promoting safety in my squadron.	52	55	49	44
	Leaders/supervisors in my squadron are successful in communicating safety goals to unit personnel.	53	56	50	
	Leaders micromanage routine operations.	54	57	51	43
	Operations Control Centers (e.g., MOC, vehicle dispatch, MUNS Control, Security Control, etc.) are effective in managing work actions for my squadron.		58		
	Work center supervisors coordinate their actions in my squadron.		59		
	Contractors are held to the same safety performance standards as military and civilian Air Force employees.		60	52	
Open-Ended Response Items	The most hazardous activity I perform is:	55	61	53	45
	The next incident/mishap in my squadron [directorate/division] will be caused by	56	62	54	46
	The most significant action(s) my squadron [directorate/division] can take to improve safety is (are):	57	63	55	47
	What is my organization [directorate/division] doing right and why?	58	64	56	48
	Use this space to provide any concern that you would like to comment upon.	59	65	57	49

Bibliography

Abalo, Carlos, "The Management Internal Control Toolset—New System Helps Streamline Program Oversight," *Citizen Airman*, April 1, 2009, p. 22.

Adam, E., "Accident Causation and the Management Systems," *Professional Safety*, 1976.

Air Force Instruction 90-201, *Inspector General Activities*, June 2009.

Air Force Instruction 38-101, *Air Force Organization*, March 2012a.

Air Force Instruction 90-201, *Inspector General Activities*, Special Management Series, March 2012b.

Air Force Manpower Agency, "2010 Air Force Climate Survey Frequently Asked Questions," undated. As of October 28, 2011:
http://www.31fss.com/Main%20Page/2010-AF-Climate-Survey-FAQ.pdf

———, *2003 Air Force Climate Survey*, Randolph Air Force Base, Tex., 2002.

———, "Air Force Climate Survey Yields Insights," Air Force Personnel, Services and Manpower Public Affairs, Randolph Air Force Base, Tex., April 20, 2011.

Air Force Policy Directive 36-29, *Military Standards*, October 29, 2009.

Air Force Safety Center, *Operations Survey, Maintenance Survey, Support Survey, and Higher Headquarters Survey*, Kirtland Air Force Base, N.M.: AFCAST Program, undated.

Alberts, David S., and Richard E. Hayes, *Understanding Command and Control*, Command and Control Research Program, Office of the Assistant Secretary of Defense (Networks and Information Integration), Department of Defense, Washington, D.C., 2006.

Anderson, Shanon E., "A MICT User's Perspective and Demonstration," briefing, 305th Air Mobility Wing, Joint Base McGuire-Dix-Lakehurst, N.J., February 2011.

———, *Employing the Management Internal Control Toolset (MICT) Across the Enterprise*, graduate research project, Air Force Institute of Technology, Wright-Patterson Air Force Base, Ohio, AFIT/IMO/ENS/12-02, May 2012.

Arrow, Kenneth J., *Limits of Organization*, New York: Basic Books, 1974.

Avolio, B. J., and B. M. Bass, *Manual for the Multifactor Leadership Questionnaire (Form 5X)*, Redwood City, Calif.: Mind Garden, 2002.

Bartis, James T., Frank Camm, and David Santana Ortiz, *Producing Liquid Fuels from Coal: Prospects and Policy Issues*, Santa Monica, Calif.: RAND Corporation, MG-754-AF/NETL, 2008. As of February 6, 2013: http://www.rand.org/pubs/monographs/MG754.html

Bass, Bernard M., *Leadership and Performance Beyond Expectations*, New York: Free Press, 1985.

Bass, Bernard M., and Bruce J. Avolio, *Full Range Leadership Development: Manual for the Multifactor Leadership Questionnaire*, Menlo Park, Calif.: Mind Garden, 1997.

Bass, Bernard M., Bruce J. Avolio, Dong I. Jung, and Yair Berson, "Predicting Unit Performance by Assessing Transformational and Transactional Leadership," *Journal of Applied Psychology*, Vol. 88, 2003, pp. 207–218.

Bikson, Tora K., and J. D. Eveland, "Groupware Implementation: Reinvention in the Sociotechnical Frame," Santa Monica, Calif.: RAND Corporation, RP-703, 1998. As of February 6, 2013: http://www.rand.org/pubs/reprints/RP703.html

Bird, F., and R. Loftus, *Loss Control Management*, Loganville, Ga.: Institute Press, 1976.

Camm, Frank, "Adapting Best Commercial Practices to Defense," in Stuart E. Johnson, Martin C. Libicki, and Gregory F. Teverton, eds., *New Challenges, New Tools for Defense Decisionmaking*, Santa Monica, Calif.: RAND Corporation, MR-1576-RC, 2003, pp. 211–246. As of February 6, 2013: http://www.rand.org/pubs/monograph_reports/MR1576.html

Camm, Frank, et al., *Implementing Proactive Environmental Management: Lessons Learned from Best Commercial Practice*, Santa Monica, Calif.: RAND Corporation, MR-1371-OSD, 2001. As of February 6, 2013: http://www.rand.org/pubs/monograph_reports/MR1371.html

Ciavarelli, A. P., "Assessing Safety Climate and Culture: From Aviation to Medicine," briefing presented at the Safety Across High-Consequence Industries Conference, St. Louis, Mo., September 2005.

———, "Assessing Safety Climate and Organizational Risk," briefing presented at the Human Factors and Ergonomics Society 51st Annual Meeting, October 2007a.

———, "Safety Climate and Risk Culture: How Does Your Organization Measure Up?" *Human Factors Associates*, 2007b.

Ciavarelli, A. P., and Jeffrey Crowson, "Organizational Factors in Accident Risk Assessment," briefing presented at the Safety Across High-Consequence Industries Conference, St. Louis, Mo., March 9–10, 2004.

Ciavarelli, A. P., and R. Figlock, "Organizational Factors in Naval Aviation Accidents, in *Proceedings of the International Aviation Psychologists Conference,* Columbus, Ohio, 1997.

Ciavarelli, A. P., R. Figlock, K. Sengupta, and K. H. Roberts, "Assessing Organizational Safety Risk Using Questionnaire Survey Methods," briefing presented at the 11th International Symposium on Aviation Psychology, Columbus, Ohio, March 2001.

Cook, Cynthia R., Laura Werber Castaneda, and Abigail Haddad, "Implementation," in *Sexual Orientation and U.S. Military Personnel Policy: An Update of RAND's 1993 Study,* Santa Monica, Calif.: RAND Corporation, MG-1056-OSD, 2010, pp. 371–388. As of February 6, 2013: http://www.rand.org/pubs/monographs/MG1056.html

Creech, Bill, *The Five Pillars of TQM: How to Make Total Quality Management Work for You*, New York: Truman Talley Books/Plume, 1994.

Curphy, G. J., "An Empirical Investigation of the Effects of Transformational and Transactional Leadership on Organizational Climate, Attrition, and Performance," in K. E. Clark, M. B. Clark, and D. R. Campbell, eds., *Impact of Leadership*, Greensboro, N.C.: The Center for Creative Leadership, 1992, pp. 177–187.

Curry, Rich, "Self Inspection Back to the Future," Tinker Air Force Base, Okla.: 507th Air Refueling Wing, January 14, 2009.

Department of Defense Human Factors Analysis and Classification System (DoD HFACS), *A Mishap Investigation and Data Analysis Tool*, undated.

Desai, Vinit M., K. H. Roberts, and A. P. Ciavarelli, "The Relationship Between Safety Climate and Recent Accidents: Behavioral Learning and Cognitive Attributions," *Human Factors,* Vol. 48, 2006, pp. 639–650.

Dvir, Taly, Dov Eden, Bruce J. Avolio, and Boas Shamir, "Impact of Transformational Leadership on Follower Development and Performance: A Field Experiment," *Academy of Management Journal*, Vol. 45, No. 4, 2002, pp. 735–744.

Edmondson, Amy C., "Learning from Mistakes Is Easier Said Than Done: Group and Organizational Influences on the Detection and Correction of Human Error," *Journal of Applied Behavioral Science*, Vol. 32, 1996, pp. 5–32.

———, "Psychological Safety and Learning Behavior in Work Teams," *Administrative Science Quarterly*, Vol. 44, No. 2, 1999, pp. 350–383.

———, "Strategies for Learning from Failure," *Harvard Business Review*, April 2011, pp. 48–55.

Edwards, E., "Introductory Overview," in E. Wiener and D. Nigel, eds., *Human Factors in Aviation*, San Diego, Calif.: Elsevier, 1988, pp. 3–25.

Federal Aviation Administration, *Crew Resource Management Training*, AC 120-51B, Washington, D.C., 1997, p. 2.

———, "Air Transportation Oversight System," Vol. 10 of *Flight Standards Information Management System*, Order 8900.1, September 13, 2007.

Federal Aviation Administration Order 8900.1, *Flight Standards Information Management System*, October 2010.

Fernandez, Sergio, and Hal G. Rainey, "Managing Successful Organizational Change in the Public Sector," *Public Administration Review*, Vol. 66, No. 2, 2006, pp. 168–176.

Gaba, David M., Sara J. Singer, Anna D. Sinaiko, Jennie D. Bowen, and Anthony P. Ciavarelli, "Differences in Safety Climate Between Hospital Personnel and Naval Aviators," *Human Factors,* Vol. 45, 2003, pp. 173–185.

Galpin, Timothy J., *The Human Side of Change: A Practical Guide to Organization Redesign*, 1st ed., San Francisco: Jossey-Bass Publishers, 1996.

Hackman, J. Richard, and Ruth Wageman, "Asking the Right Questions About Leadership," *American Psychologist*, Vol. 62, No. 1, 2007, pp. 43–47.

Heinrich, H., D. Peterson, and N. Roos, *Industrial Accident Prevention*, 5th ed., New York: McGraw-Hill, 1980.

Helmreich, R., and H. Foushee, "Why Crew Resource Management? Empirical and Theoretical Bases of Human Factors Training in Aviation," in E. Wiener, B. Kanki, and R. Helmreich, eds., *Cockpit Resource Management,* San Diego, Calif.: Elsevier, 1993, pp. 3–45.

Hiller, Nathan J., Leslie A. DeChurch, Toshio Murase, and Daniel Doty, "Searching for Outcomes of Leadership: A 25-Year Review," *Journal of Management*, Vol. 37, No. 4, 2011, pp. 1137–1177.

Hyde, Robert, "Chapter 6—Commander's Inspection Program (CCIP)," draft version 4, September 2011a.

———, "Inspection System Improvement Tiger Team (ISITT) Efforts," briefing presented at Air Force Inspector General Conference, Kirtland Air Force Base, N.M., October 3, 2011b.

Jensen, Michael C., *Foundations of Organizational Strategy*, Cambridge, Mass.: Harvard University Press, 1998.

Judge, Timothy A., and Ronald F. Piccolo, "Transformational and Transactional Leadership: A Meta-Analytic Test of Their Relative Validity," *Journal of Applied Psychology*, Vol. 89, No. 5, 2004, pp. 755–768.

Judson, A. S., *Changing Behavior in Organizations: Minimizing Resistance to Change*, Cambridge, Mass.: Basil Blackwell, 1991.

Kaiser, Robert B., Robert Hogan, and S. Bartholomew Craig, "Leadership and the Fate of Organizations," *American Psychologist,* Vol. 63, No. 2, 2008, pp. 96–110.

Khanna, Madhu, and Diah Widyawati, "Fostering Regulatory Compliance: The Role of Environmental Self-Auditing and Audit Policies," *Review of Law and Economics*, Vol. 7, 2011, pp. 29–164.

Kotter, John P., *Leading Change*, Boston: Harvard Business School Press, 1996.

Libuser, C. B., *Organizational Structure and Risk Mitigation*, Ph.D. dissertation, Los Angeles, Calif.: University of California at Los Angeles, 1994.

Lieberson, Stanley, and James F. O'Connor, "Leadership and Organizational Performance: A Study of Large Corporations," *American Sociological Review*, Vol. 37, No. 2, 1972, pp. 117–130.

Mackey, Alison, "The Effect of CEOs on Firm Performance," *Strategic Management Journal*, Vol. 29, 2008, pp. 1357–1367.

March, James G., *A Primer of Decision Making: How Decisions Happen*, New York: Free Press, 1994.

Miles, Matthew B., and Michael Huberman, *Qualitative Data Analysis: An Expanded Sourcebook,* 2nd ed., Thousand Oaks, Calif.: Sage Publications, 1994.

Moore, Nancy Y., Laura H. Baldwin, Frank Camm, and Cynthia R. Cook, *Implementing Best Purchasing and Supply Management Practices: Lessons from Innovative Commercial Firms,* Santa Monica, Calif.: RAND Corporation, DB-334-AF, 2002. As of February 6, 2013: http://www.rand.org/pubs/documented_briefings/DB334.html

National Transportation Safety Board, "Aircraft Accident Report: Uncontrolled Collision with Terrain. American International Airways Flight 808, Douglas DC-8-61, N814CK, U.S. Naval Station, Guantanamo Bay, Cuba, August 18, 1995," NTSB/AAR-94-04, Washington, D.C., 1994.

O'Hare, D., M. Wiggins, R. Batt, and D. Morrison, "Cognitive Failure Analysis for Aircraft Accident Investigation," *Ergonomics,* Vol. 37, 1994, pp. 1855–1869.

O'Reilly, Charles A., David F. Caldwell, Jennifer A. Chatman, Margaret Lapiz, and William Self, "How Leadership Matters: The Effects of Leaders' Alignment on Strategy Implementation," *The Leadership Quarterly,* Vol. 21, 2010, pp. 104–113.

Patterson, J., and S. Shappell, "Operator Error and System Deficiencies: Analysis of 508 Mining Incidents and Accidents from Queensland, Australia, Using HFACS," *Accident Analysis and Prevention,* Vol. 42, 2010, pp. 1379–1385.

Rasmussen, J., "Human Errors: A Taxonomy for Describing Human Malfunction in Industrial Installations," *Journal of Occupational Accidents,* Vol. 4, 1982, pp. 311–333.

Reason, J., *Human Error,* New York: Cambridge University Press, 1990.

———, *Managing the Risks of Organizational Accidents,* Brookfield, Vt.: Ashgate, 1997.

Reinhart, R., *Human Error,* New York: Cambridge University Press, 1996.

Roberts, Karlene, "Managing High Reliability Organizations," *California Management Review,* Vol. 32, No. 4, 1990, pp. 101–113.

———, "Culture Characteristics of Reliability Enhancing Organizations," *Journal of Managerial Issues,* Vol. 5, 1993, pp. 165–181.

Roberts, K., D. Rousseau, and T. La Porte, "The Culture of High Reliability: Quantitative and Qualitative Assessment Aboard Nuclear Powered Aircraft Carriers," *Journal of High Technology Management Research,* Vol. 5, 1994, pp. 141–161.

Rogers, Marc E., "Inspection/Assessment Frequency," briefing to CORONA Top, SAF/IG, Washington, D.C., 2010.

———, "Opening Remarks to Air Force Inspector General Conference," briefing presented at Air Force Inspector General Conference, Kirtland Air Force Base, N.M., October 3, 2011.

Salancik, Gerald R., and Jeffrey Pfeffer, "Constraints on Administrator Discretion: The Limited Influence of Mayors on City Budgets," *Urban Affairs Quarterly,* Vol. 12, 1977, pp. 475–498.

Salomon, Richard, "Air Force Climate Survey: How Do You Really Feel? Air Force Climate Survey Offers Chance to 'Speak Today, Shape Tomorrow,'" *Airman,* October 2003.

Schein, Edgar H., "How Can Organizations Learn Faster? The Challenge of Entering the Green Room," *Sloan Management Review,* Winter 1993, pp. 85–92.

Shappell, S., and D. Wiegmann, "U.S. Naval Aviation Mishaps 1977–1992: Differences Between Single- and Dual-Piloted Aircraft," *Aviation, Space, and Environmental Medicine,* Vol. 67, 1996, pp. 65–69.

———, "The Human Factors Analysis and Classification System—HFACS," U.S. Department of Transportation Federal Aviation Administration Office of Aviation Medicine, DOT/FAA/AM-00/7, 2000.

———, "Unraveling the Mystery of General Aviation Controlled Flight into Terrain Accidents Using HFACS," presented at the 11th International Symposium on Aviation Psychology, Ohio State University, Columbus, Ohio, 2001.

Smith, Debi, "Sheppard Innovates Web-Based Self-Inspections," Sheppard Air Force Base, Tex.: Air Force News Service, August 5, 2011.

Stecher, Brian M., Frank Camm, Cheryl L. Damberg, Laura S. Hamilton, Kathleen J. Mullen, Christopher Nelson, Paul Sorensen, Martin Wachs, Allison Yoh, Gail L. Zellman, and Kristin J. Leuschner, *Toward a Culture of Consequences: Performance-Based Accountability Systems for Public Services*, Santa Monica, Calif.: RAND Corporation, MG-1019, 2010. As of February 6, 2013: http://www.rand.org/pubs/monographs/MG1019.html

Wang, Gang, In-Sue Oh, Stephen H. Courtright, and Amy E. Colbert, "Transformational Leadership and Performance Across Criteria and Levels: A Meta-Analytic Review of 25 Years of Research," *Group and Organization Management*, Vol. 36, No. 2, 2011, pp. 223–270.

Weaver, D., "Symptoms of Operational Error," *Professional Safety*, 1971.

Weiner, N., and T. A. Mahoney, "A Model of Corporate Performance as a Function of Environmental, Organizational, and Leadership Influences," *Academy of Management Journal*, Vol. 24, 1981, pp. 453–470.

Wiegmann, D., A. Rich, and S. Shappell, *Human Error and Accident Causation Theories Frameworks and Analytical Techniques: An Annotated Bibliography*, Savoy, Ill.: University of Illinois Aviation Research Lab, Technical Report ARL-00-12/FAA-00-7, 2000.

Wiegmann, D., and S. Shappell, "Human Factors Analysis of Post-Accident Data: Applying Theoretical Taxonomies of Human Error," *The International Journal of Aviation Psychology*, Vol. 7, 1987, pp. 67–81.

———, "Human Error and Crew Resource Management Failures in Naval Aviation Mishaps: A Review of U.S. Naval Safety Center Data, 1990–1996," *Aviation, Space, and Environmental Medicine*, Vol. 70, 1999, pp. 1147–1151.

———, "Human Errors Perspectives in Aviation," *The International Journal of Aviation Psychology*, Vol. 11, No. 4, 2000, pp. 341–357.

———, "Applying the Human Factors Analysis and Classification System (HFACS) to the Analysis of Commercial Aviation Accident Data," presented at the 11th International Symposium on Aviation Psychology, Ohio State University, Columbus, Ohio, 2001.

Zellman, Gail L., et al., "Implementing Policy in Large Organizations," in *Sexual Orientation and U.S. Military Personnel Policy: Options and Assessment*, Santa Monica, Calif.: RAND Corporation, MR-323-OSD, 1993, pp. 368–394. As of February 6, 2013: http://www.rand.org/pubs/monograph_reports/MR323.html

Zohar, Dov, "The Effects of Leadership Dimensions, Safety Climate, and Assigned Priorities on Minor Injuries in Work Groups," *Journal of Organizational Behavior*, Vol. 23, 2002, pp. 75–92.